Mapping Landscapes for Performance as Research

Also By Lynette Hunter

CRITIQUES OF KNOWING: Situated Textualities in Science, Computing and the Arts

LITERARY VALUE AND CULTURAL POWER

OUTSIDER NOTES: Feminist Approaches to Canadian Publishing, Writers and Readers, 1960–1990

SHAKESPEARE, LANGUAGE AND THE STAGE: The Fifth Wall, Issues in Practice and Criticism *(edited with Peter Lichtenfels)*

Mapping Landscapes for Performance as Research

Scholarly Acts and Creative Cartographies

Edited by

Shannon Rose Riley and Lynette Hunter

palgrave
macmillan

First published 2009 by
PALGRAVE MACMILLAN

Palgrave Macmillan in the UK is an imprint of Macmillan Publishers Limited, registered in England, company number 785998, of Houndmills, Basingstoke, Hampshire RG21 6XS.

Palgrave Macmillan in the US is a division of St Martin's Press LLC, 175 Fifth Avenue, New York, NY 10010.

Palgrave Macmillan is the global academic imprint of the above companies and has companies and representatives throughout the world.

Palgrave® and Macmillan® are registered trademarks in the United States, the United Kingdom, Europe and other countries.

ISBN-13: 978–0–230–22219–9 hardback

This book is printed on paper suitable for recycling and made from fully managed and sustained forest sources. Logging, pulping and manufacturing processes are expected to conform to the environmental regulations of the country of origin.

A catalogue record for this book is available from the British Library.

A catalog record for this book is available from the Library of Congress.

10 9 8 7 6 5 4 3 2 1
18 17 16 15 14 13 12 11 10 09

Printed and bound in Great Britain by
CPI Antony Rowe, Chippenham and Eastbourne

Contents

List of Figures

Acknowledgements

The editors and publishers wish to thank the following for permission to reproduce copyright material:

Bob Raymond for his photograph of *Reports from the Interior* by Shannon Rose Riley (1999), performed at Mobius, Inc., Boston.

Paul Clarke and Ben Pacey for Pacey's photograph of "Apple Bobbing Scene" in *It is Like it Ought to Be: A Pastoral* by Uninvited Guests (2006), Nuffield Theatre, Lancaster, UK.

Drew Yapp for his photograph of *Being in Between* co-directed by Baz Kershaw and Sandra Reeve (2005) Bristol Zoological Gardens, Bristol, UK.

Simon Jones for Edward Dimsdale's photograph of *Model Love* by Bodies in Flight (2008) Arnolfini, Bristol, UK.

Paul Clarke for Tom Hall's photograph of *Offline* by Uninvited Guests (2001) Arnolfini, Bristol, UK.

Arnolfini for Carl Newland's photograph of *Niceday* by Shi Ker and Steve Robins (2007) Arnolfini, Bristol, UK.

Mary Gearhart for her photograph of *To You, The Birdie! (Phèdre)* directed by Elizabeth LeCompte.

Sofia De Grenade for her photograph of *undertow*, a durational performance by Marilyn Arsem, in Ex-Frigorífico Empresa Portuaria Valparaíso at POST, at the 7ma Muestra internacional de Arte de Performance, 19 November 2005, as part of the 1er Congreso Internacional de Arte de Performance, Valparaíso, Chile, presented by PerfoPuerto.org.

William Swartout, Director of the Institute of Creative Technologies, University of Southern California for the photograph of the "holodeck" at the Institute for Creative Technologies (ICT) at the University of Southern California.

Mark Cetilia for his photograph of Joseph Butch Rovan in performance with MiMICS (2007) at NIME, New York.

South African Theatre Journal (SATJ) for permission to quote from Temple Hauptfleisch, "Artistic Outputs, Arts Research and the Rating of the Theatre Practitioner as Researcher – Some Responses to the NRF Rating

System after the First Three Years," in *South African Theatre Journal* (2005) volume 19.

Theatre Academy for pp. 118–23 from Annette Arlander, "Some Conversations [. . .] in Various Spaces," in *Knowledge Is a Matter of Doing: Proceedings of the Symposium Theatre and Dance Artists Doing Research in Practice*, Acta Scenica 1, ed. P. Paavolainen and A. Ala-Korpela, Helsinki, Theatre Academy (1995).

Theatre Academy for pp. 48–65 from Annette Arlander, "Tila Ja Aika Eli Miten Megalomaanisesta Idealistista Tuli Suhteellisuudentajuinen Pragmaatikko (Space and Time – or How a Megalomaniac Idealist Turned into a Pragmatist with a Sense of Proportion)," in *Taide, Kertomus Ja Identiteetti (Art, Narrative and Identity)*, Acta Scenica 3, ed. P. Houni and P. Paavolainen, Helsinki, Theatre Academy (1999)

Every effort has been made to trace rights holders, but if any have been inadvertently overlooked the publishers would be pleased to make the necessary arrangements at the first opportunity.

The editors would also like to thank the members of the Performance as Research Seminar convened by Kris Salata and Lisa Wolford Wylam at the 2006 American Society of Theatre Research (ASTR) conference, Chicago. Several of the contributors in this collection were part of that group. We are enormously grateful to all of the contributors for the gift of their time and insight and for putting up with endless emails and edits – and to Sylvie Bissonnette for her assistance in reading the proofs and her thoughts on the index. Do Mi Stauber completed the final index with very little notice – we are especially thankful for her great skill. We very much appreciate the financial support and editorial time provided by the Center for Women's Intercultural Leadership at Saint Mary's College, Notre Dame; the University of California, Davis; and San José State University. We might have lost our minds were it not for the keen eye and delightfully sharp wit of our talented copy editor, Penny Simmons. We would especially like to thank our families for putting up with our hectic schedules – endlessly.

Abbreviations and Acronyms

AHRC	Arts and Humanities Research Council (UK)
CAFAD	Canadian Association of Fine Arts Deans
CPR	Centre for Performance Research (Wales)
NRF	National Research Foundation (South Africa)
OECD	Organisation for Economic Co-Operation and Development
PbR or **PBR**	Practice-based research
PaR	Practice as research
PAR	Performance as research
RAE	Research Assessment Exercise (UK)
SSHRC	Social Sciences and Humanities Research Council (Canada)

Other acronyms for organizations, institutions, and universities are explained in individual essays, as needed.

Notes on Contributors

Annette Arlander is Professor of Performance Art and Theory and head of the Performing Arts Research Centre at the Theatre Academy, Finland.

Marilyn Arsem is Regular Full-time Faculty and the head of the Performance Area at the School of the Museum of Fine Arts, Boston. She is also the founder of Mobius, Inc., a Boston-based collaborative of interdisciplinary artists.

Sharon Marie Carnicke, Professor of Theatre and Slavic Languages and Literatures at the University of Southern California, is the author of *Stanislavsky in Focus* (Second Edition, 2008).

James Elkins is Professor of Art History, Theory, and Criticism at the School of the Art Institute of Chicago.

Faye Chunfang Fei is Professor of English and Drama, and Director of American Studies at East China Normal University in Shanghai.

Susan Leigh Foster is Professor in the Department of World Arts and Cultures at the University of California, Los Angeles.

Richard Gough is Professor at Aberystwyth University, Artistic Director of the Centre for Performance Research and General Editor of *Performance Research*.

Sandra Jean Graham is Assistant Professor in the Department of Music at the University of California, Davis.

Brad Haseman is Professor and Assistant Dean (Research) for the Creative Industries Faculty at the Queensland University of Technology, Australia.

Temple Hauptfleisch is Director of the Centre for Theatre and Performance Studies at Stellenbosch. Formerly head of the Centre for SA Theatre Research and Chair of the University of Stellenbosch Drama Department, he is the co-founder and current editor of *South African Theatre Journal*.

Lynette Hunter is Professor of the History of Rhetoric and Performance in the Department of Theatre and Dance at the University of California, Davis.

Shannon Jackson is Professor in the Departments of Rhetoric and Theater, Dance and Performance Studies at the University of California, Berkeley.

Simon Jones is Professor of Performance at the University of Bristol and Co-Director of the performance company Bodies in Flight.

Nick Kaye is Chair in Performance Studies, University of Exeter. He was formerly Chair in Drama, University of Manchester.

Susan Kelly is Lecturer in Fine Art at Goldsmiths College, University of London.

Baz Kershaw is Professor of Performance at the University of Warwick, formerly Chair of Drama at the University of Bristol and director of PARIP, a five-year project investigating Practice as Research in Performance.

Laura Levin is Assistant Professor in the Department of Theatre at York University.

Peter Lichtenfels is Professor in the Department of Theatre and Dance at the University of California, Davis.

Rosemary Malague is Director of the Theatre Arts Program and Senior Lecturer in Theatre Arts at the University of Pennsylvania.

Lara Nielsen is Assistant Professor in the Department of Theater and Dance at Macalester College, St Paul, Minnesota.

Ilya Noé is a PhD Candidate in the Practice-Based PhD Program in Performance Studies at the University of California, Davis.

Jane Plastow is Professor of African Theatre at the University of Leeds, Deputy Director of the Workshop Theatre, and Director of Leeds University Centre for African Studies.

Della Pollock is Professor of Performance and Cultural Studies in the Department of Communication Studies at the University of North Carolina, Chapel Hill.

Shannon Rose Riley is Assistant Professor of Humanities and Coordinator of the Creative Arts Program at San José State University.

Jon D. Rossini is Associate Professor in the Department of Theatre and Dance at the University of California, Davis.

Joseph Butch Rovan is Associate Professor in the Department of Music at Brown University and co-director of MEME: Multimedia and Electronic Music Experiments at Brown.

Arthur J. Sabatini is Associate Professor of Performance Studies in the Department of Interdisciplinary Arts and Performance at Arizona State University. He is presently working on a book, *Aesthetic Research in the 20th/21st Century: History, Philosophy, Practice*.

Henry Spiller is Assistant Professor in the Department of Music at the University of California, Davis and a long-time student, performer, and teacher of Sundanese and Javanese gamelan music.

William Huizhu Sun is Professor and Vice-President at Shanghai Theatre Academy and a contributing editor of *TDR*.

John T. Warren is Associate Professor of Speech Communication at Southern Illinois University, Carbondale.

Ian Watson is the Chair of the Department of Visual and Performing Arts and Professor and Coordinator of the Theatre Program at Rutgers University, Newark.

Julie Wyman is Assistant Professor of Technocultural Studies at the University of California, Davis.

Kim Yasuda is Professor in the Department of Art at the University of California, Santa Barbara and Co-Director of the University of California Institute for Research in the Arts.

Introduction

Although the sciences have long understood the value of practice-based research, the modern arts and humanities have tended to structure a gap between practice and analysis. While performance practices have always contributed to knowledge, the idea that performance can be more than creative production, that it can constitute intellectual inquiry and contribute new understanding and insight is a concept that challenges many institutional structures and calls into question what gets valued as knowledge. Perhaps the most singular contribution of the developing areas of practice as research (PaR) and performance as research (PAR) is the claim that creative production can constitute intellectual inquiry.

The chapters in this book map out a landscape, always partial, of PAR in the United States in a way that both acknowledges the legacies and influences of PaR elsewhere and pays attention to the particular influences of the way that performance studies has developed in the United States. The intention is to delineate PAR in the United States more clearly, and in so doing contribute to the discussions around PaR and PARIP (Practice as Research in Performance) in the United Kingdom, Europe, and elsewhere in the world. The chapters in the first section include perspectives for understanding the interrelated but different developments of PaR and PAR within different geographical and institutional contexts, and for comparing the institutionalization of PAR with similar programs in PaR in the arts and new humanities. The central section offers a series of chapters on research terms that provide the reader with a toolkit of cartographies, or ways of continuing to map out and produce PAR. The collection moves on to provide examples of PAR being used in a broad range of courses in the humanities, from music to theatre, from the visual arts to performance art, new media, and technology, and concludes with a series of commentaries on the scholarly interactions and contributions of practice as research productions. The editors aim for the collection to prove useful in generating continued research projects on PAR and its development in different contexts.

PaR and PAR (see below for an expanded discussion) constitute part of a revolution in how we look at knowledge today, and there has been a boom in such research approaches in the arts and humanities over the last two decades – predominantly in the United Kingdom and some parts of Europe but increasingly elsewhere. The United States, having lagged behind partly as a result of the varying institutional and economic structures of higher education in different countries – for example, public funding is perhaps more directly responsible for institutional imperatives to define and produce "research" in the United Kingdom, Australia, and South Africa than in the United States – is, as this book demonstrates, becoming a significant contributor to this field of knowledge.

The goal of this collection is to bring the United States into dialogue with work in practice as research in performance and the arts in general. The development and usage of "practice as research" in the United Kingdom and elsewhere is not exactly the same as the concept and practice of "performance as research" in the United States. There are many areas of overlap, but the differences between them are quite instructive. This collection is structured so as to point to these differences, with the chapters in Part I mapping some of the landscapes of practice as research (PaR) in the arts, predominantly in performance, in the United Kingdom, Finland, Canada, Australia, China, and South Africa. Part I also includes grounding chapters on performance as research (PAR) work in the United States: for example, James Elkins describes the three configurations of current "studio-art PhD" programs in the United States and United Kingdom; Sandra Graham writes about the emergence of performance as research in ethnomusicology in North America; and Susan Leigh Foster offers a reading of performance as research legacies in teaching dance in the United States in the early and middle twentieth century.

The chapters in Part I all provide fundamental perspectives on work in PaR and PAR in terms of disciplinary and institutional legacies. Against this background, Part III maps out contemporary case studies in research and teaching on PAR in the United States. This structure allows us to put these geographies and disciplines into conversation, encouraging them to foreground important, or as Kershaw puts it, "instructive" differences. Part II offers a cartography of points of commonality and further conversation. It includes short entries and questioning perspectives on key terms that are used frequently in the chapters in both Parts I and III, terms that are beginning to be recognized as central to the conceptual vocabulary for this new field and yet are even now also beginning to be critiqued. The book brings the United States into the conversation on PaR/PAR, maps out landscapes for PAR in relation to PaR and juxtaposes experiences with historical/institutional developments, teaching, and research practices in PaR/PAR, that begin to hint at the complexity of this new disciplinary field.

Current contexts

The title of the collection reflects prominent themes that run through many of the chapters. The main title, *Mapping Landscapes for Performance as Research,* aims primarily through the chapters in Part I to enrich the conversation on PaR/PAR by beginning to map out similarities and differences in various institutional, community, and national contexts. There is a view of PAR specifically situated in the United States and to some extent in Canada that is subtly different from the dominant view of PaR in performance as it has currently developed in the United Kingdom and other parts of the world. Differences in terminology are particularly visible at the level of the acronym.

The acronym "PaR" in the United Kingdom refers to "practice as research" in its most inclusive sense to embrace music practices, the visual arts, dance, and theatre and "PbR" refers to "practice-based research" with a wider reach across the arts and sciences.[1] PbR is also well-established in the United Kingdom and Europe, and contributes to many areas, from the medical sciences to spectatorship studies. PbR emerges from different academic areas, but seems to have particular usage in the sciences.[2] In the United States, especially in the fields of performance and theatre studies, the acronym PAR is common shorthand for "performance as research."

PaR was developed in the United Kingdom for a number of years partly as a result of the Research Assessment Exercises (RAE) brought in by the British government in the early 1990s. The RAE determines which universities and which departments will be funded for research on the basis of several indicators, including research output. Because of the humanities paradigm of written monographs, essays and articles, theatre studies that were practice-based departments – and many theatre departments are not practice-based – were excluded from assessment. Historically, arts departments of all kinds are new to the university system, and there is very little understanding of what might constitute research in the arts or artistic research or art as research. Between the 1992 and 1996 RAEs, SCUDD (Standing Conference of University Drama Departments) created a PaR Working Group, chaired by contributor Baz Kershaw, that produced an advisory report submitted to the 1996 RAE Panel for Dance, Drama and Performing Arts. This initiative developed PaR criteria that were acceptable as "research" in the eyes of later RAE. Several contributors to this volume, such as Lynette Hunter, Jane Plastow, and Peter Lichtenfels, took part in aspects of this evolution. A further stage in the UK development of PaR was heralded by the government award (from the Arts and Humanities Research Board) to the University of Bristol to develop PARIP, and to some extent to Lancaster University to develop Palatine. The former was also headed by Baz Kershaw who is now at the University of Warwick, and who founded (with Jacqueline Martin) the Practice as Research section of the International Federation of Theatre Research (FIRT).

Currently PaR is swiftly developing in the United Kingdom and in Australia, another country where it has been recognized for over a decade. In Europe it is gaining ground, for example Finland, as it has been in Canada, but US recognition has only just begun. A seminar at the American Society of Theatre Research in 2006 demonstrated that Performance as Research in North America already included performative methods in ethnography and social anthropology as well as the arts, especially experimental theatre. The seminar was reformed as the PAR Working Group in 2007. "Performance as research" is less common in the United Kingdom as an umbrella term across a range of performance studies. For example, rather than "performance as research," the acronym PARIP is used in Kershaw's project in the United Kingdom. It is clear that the terminology of PAR and PaR requires a situated mapping out in terms of geographies, institutional histories, and practices. The terms hold a slightly different valence in different geographical and institutional contexts, and do not translate as directly as one might assume. As editors, we use PAR to signify "performance as research" and PaR to signify "practice as research." However, we preserve the original language used by all contributors to maintain the complexity of our mapping, including at the level of terminology.

Putting aside questions of funding, institutional criteria for research, and other administrative issues, the "differences" between PaR and PAR are partly the result of the situated development of the field of performance studies in the United States. Barbara Kirshenblatt-Gimblett traces the development of the field within different institutional contexts and conceptual paradigms, predominantly in the United States, by describing what she called the "broad spectrum approach" engaged at New York University and the "aesthetic communication approach" of Northwestern University.[3] The Northwestern model has produced much work in the area of social practice, communications theory/practice, and rhetoric. The NYU model has produced much work in the interdisciplinary area of anthropology, theatre, and intercultural performance.

Despite the fact that the NYU model has become increasingly associated with cultural performance, Kirshenblatt-Gimblett reiterates that its roots were in experimental performance and theatre. Her third category, "ethnoscenology" at the University of Paris VIII is primarily concerned with "contemporary experimental performance."[4] This last approach most closely mirrors the work in experimental performance that is part of the UK/EU model of PaR in performance. As a result of these and more recent developments, we suggest that there is a three-pronged approach to performance studies that is largely unique to the United States and perhaps Canada.

Because of the multiple articulations of performance studies in the United States, PAR work in this context includes work in all three areas: social practices (the Northwestern model); creative ethnographies and intercultural performance (NYU); experimental theatre, which includes a wide variety of media, not only the stage (NYU and University of Paris VIII); and more recently, practice as research in performance, which is evidenced in all models, but also found independently at an increasing number of institutions such as University of California, Davis. Social practice and creative ethnography do form a part of PaR work in the United Kingdom and elsewhere, but because of the different intellectual histories in these geographic locations, terminology does not translate directly. For example, in the United Kingdom, PaR in performance has focused especially, although not exclusively, on experimental performance practice.

Although PaR/PAR is distinguished from the development of performance studies in the United States and elsewhere, there is much crossover as faculty, students, ideas, and practices move across national boundaries. William H. Sun's essay on PAR in China is particularly instructive in this regard. Trained in performance studies in the United States, he is working at the Shanghai Theatre Academy to develop a practice as research curriculum in what he calls social performance studies. US trajectories of performance studies have had some influence on other models, and UK, Australian, and other European models of PaR have had considerable impact on PAR practices in the United States. The broadly-based chapters on the development of PaR programs and research in dance, music, theatre, visual studies, and ethnomusicology are crucial for Part I, because they not only reflect the "broad spectrum" of interests and applications of performance studies, but also provide information on how other areas of the arts have defined and/or institutionalized some form of PaR and therefore might prove useful in developing PAR programs elsewhere.

The subtitle, *Scholarly Acts and Creative Cartographies*, underlines the themes of Part II "Cartographies" and of Part III "Mapping PAR in the United States." "Cartographies" provides definitions for many of the terms that appear frequently throughout the collection; Kershaw's entry in this section describes them as "emergent keywords" in the field. "Scholarly Acts" intends to foreground the complexity of what it might mean to act in a scholarly way in the field of performance: not only to perform the role of the scholar in a particular institutional setting, but also to question how academic identities/roles are highly performative. Part III consists of seven opening chapters on "Pedagogies," and another seven on "Current Research Practices." The combination of action and scholarship in the phrase "scholarly acts" foregrounds that PAR research requires action or acting in some fashion. Many PAR research methods, such as collaboration, action research, oral history, experimental theatre and dance, constitute such acts. Because of the only recent inclusion of PaR and PAR in the academic institution, current pedagogies constitute major areas of research, but there is an additional factor: the integration of practice, research, and performance that this field investigates can mean that in particular situations, the practice of the teaching becomes action research. Differentiating the practice of teaching as research from the teaching of practice is a key concern of several chapters.

The emphasis on mapping, landscape, and cartography is meant to invoke the idea of mapping sites and spaces of current research in practice and performance. It is a recognition that, as editors, we are mapping "some" landscapes for PAR – but certainly not all. Further, many of the chapters throughout the book are interested in locations and locating, overlapping geographies, and a persistent investigation of ecologies and ecosystems, sites and spaces – from Kershaw's "small island" perspective and his keyword entry for "environment," to Nielsen's "institutional ecosystems" and "ecosystems of living histories" and the keyword entries on "site-particular" and "lab/studio." There is the sense that the writers, most of whom are scholars and practitioners, are themselves mapping out PAR landscapes. For example, Laura Levin's essay speaks of "locating the artist-researcher" in "shifting sites of research." For Levin, the artist-researcher in Canada can be found at the intersection of urban studies and performance – a kind of urban mapping through performance. In the Toronto context in particular, a number of centers for urban research have been established, which aim to explore issues of civic and public space through site-specific performance projects. Arlander, writing on her work in Finland, repositions her location as researcher along a range of points, "from space to place to landscape" and Rose Malague's students "search" for Spalding Gray at the Edinburgh Fringe Festival.

The title of this book foregrounds, as does each chapter from its situated position, the performance structures of all research: as acts performed in particular locations and as highly performative acts that attempt to call forth the very thing they are invoking. By connecting acts and landscapes, we hope to show that *how* we know what we know is partly dependent on *where* we know. The focus on location and situated knowledge is part of a more general response in the scholarly world at the moment to thinking about place, space, and geography, vital to the emerging intersections of the local, regional, national and global. This book aims to provide information to scholars and artists interested in PAR, specifically as it has developed in the United States, but within a larger context that engages with the global discussions on both practice and performance as ways of creating new insights and contributing to transcultural knowledge and understanding.

Part I Lay of the Landscape – Geographies, Legacies, Disciplines

The first part of this book includes chapters on the geographies and legacies of the development and institutionalization of Practice as Research (PaR) in dance, music, theatre, and the studio arts in the United Kingdom, Australia, Canada, China, Finland, Ethiopia/Sudan/Eritrea, and South Africa. These chapters explore the ways that practice as research (PaR) has been developed in different national and institutional contexts and within different disciplines in the arts as well as the conditions under which it has developed in the academy. This offers a partial mapping, or the lay of the land, of PaR that provides a context for the development of performance as research (PAR) in the United States. This lay of the land provokes "instructive" differences between PaR and PAR to emerge in relationship to each other.

Baz Kershaw's chapter, "Practice as Research: Perspectives from a Small Island" opens the collection by mapping a lay of the land of "practice as research" in theatre and performance, predominantly as it has developed in the United Kingdom. Richard Gough's chapter discusses performance as a valued research area in the Welsh public and academic world, and provides contexts from other parts of the world for this phenomenon. He investigates the artistic and institutional background to changing attitudes to performance, and to the valuation of particular regional knowledge in a global context. Simon Jones writes on the institutional interaction of the major academic and arts funding bodies in the United Kingdom, and their joint impact on practice as research and the arts community. Jane Plastow looks at a variety of projects using theatre as a tool of practice as research in the Horn of Africa – in Eritrea, Ethiopia, and the Sudan – as well as how this work is brought into teaching in the United Kingdom. She asks how this work might help in the recognition of PaR in the United Kingdom. Temple Hauptfleisch argues that "The fight to recognize practice as research (PaR) is an international phenomenon waged on many fronts and the core issues have become points of spirited debate internationally." His work focuses on practice as research assessment in South Africa, analyzing the fallout of what he describes as the "new, highly pragmatic outcomes-based (or rather *income*-based), approach to tertiary education by the ANC-led government." Brad Haseman offers an "Australian Story" of the development of methodologies and epistemologies around practice as research and of the legitimating institutional strategies, yet crucially, also warns of the difficulties in balancing a change of perspective toward scholarly work with the construction of a rigorous evaluation for these new modes of production. Annette Arlander describes her graduate work at the Theatre Academy in Helsinki, Finland, where she was one of the first to complete a doctoral project in "artistic research," and its subsequent development, which she describes as moving along an arc "from space to place to landscape." Laura Levin describes the historical and geographic particularities of "practice as research" in Canada, focusing on performance in Ontario as an illuminating case study and William Huizhu Sun and Faye Chunfang Fei talk about the pragmatic turn in performance studies in China, as it justifies its place in higher education in the modern world.

The remaining chapters in Part I provide a partial mapping of the spaces created through disciplinary legacies and institutional boundaries. Whereas the previously described chapters in Part I offer readings on the development of practice as research in performance in various geographic settings, these investigate the legacies of the division between aesthetic or performance knowledge and "scientific" knowledge, the different legacies of performance as research in theatre, dance, and music, and the institutionalization of "Studio-Art PhDs" in the United Kingdom and United States. Ian Watson and Susan Leigh Foster each analyze specific legacies of performance as research in theatre and dance respectively. Watson traces the practice as research legacies of Stanislavsky, Grotowski, and Barba in acting training in the United States, particularly in terms of parsing out training from research and analyzing the transmission of embodied knowledge. Foster examines three distinctive US approaches developed in the 1930s, 1950s and 1960s – by Louis Horst, Daniel Nagrin, and Robert Ellis Dunn. Foster conducts a comparative analysis of their approaches to teaching choreography in order to elaborate distinctive methodologies for undertaking practice as research in dance and movement. Sandra Jean Graham writes on the development of performance as research in ethnomusicology, predominantly in the United States, but also in Canada, and in his chapter, "The Three Configurations of Studio-Art PhDs," James Elkins describes the three forms of Studio-Art PhD programs as they are currently being developed in the United States and the United Kingdom, arguing that each of the models could be implemented in a number of ways. As the last contribution to the lay of the land section, Arthur J. Sabatini's chapter traces out the parameters of the aesthetic/performance vs. "scientific" knowledge debate in modern philosophy.

Part I offers a sense of how PaR has developed in theatre and performance in different locations, how it has developed in music and dance and studio art, highlighting the relationship between visual arts and performance art. It builds an understanding of both the institutional and disciplinary development of PaR, especially in performance, and offers some background on the modern philosophy of knowledge by which we quantify and qualify what counts as research.

Part II Cartographies

Part II includes nine entries for emergent keywords that appear frequently in the collection. The entries also contain useful bibliographical information for further reading and give the reader a basic toolkit for studying, conducting, and implementing PAR projects. The terms can be theorized as belonging to two different groups. One is a collection of types of scholarly acts in performance as research, such as action research, collaboration, oral history, and ways of knowing that are embodied and situated. The other terms constitute a basic terminology for mapping spaces and sites of research: "environment," "medium," "lab/studio," "site specific/site particular," and "media space" (disjunction). Some terms belong to both groups: for example, "embodied knowledge" is both a kind of research act (embodied knowing) and a kind of scape or site of knowledge (bodyscape, etc.). "Situated knowledge," too, invokes both an act and process of knowing embedded inextricably in a location or situation.

Part III Mapping PAR in the United States – Communities, Classrooms, Studios, and Holodecks

Part III consists of work in two main areas: pedagogies and current research practices, and focuses on the disciplinary area of performance as research in the United States.

The first seven chapters offer a range of approaches to, and understandings of, using practice as research in teaching. Shannon Jackson opens the section with a specific study of teaching the arts as a research form at the university level in the United States. Her chapter is followed by six others that explore specific examples of using performance as research in a graduate or undergraduate classroom context. Lara D. Nielsen explores an undergraduate course that she teaches in/on oral history. Henry Spiller describes teaching music ethnography through PAR approaches, drawing on his experience leading research-based gamelan ensembles at five institutions of higher learning over the past two decades. John T. Warren provides examples of PAR work that he uses in the classroom in order to "performatively unveil whiteness, showing how normalized cultural patterns are constituted." Working from Deleuze's challenge to teaching, Kim Yasuda explores the relation between pedagogy, action research, and art. Rosemary Malague describes a course she teaches at the University of Pennsylvania, which culminates with a student performance at the Edinburgh Fringe Festival, and Lynette Hunter describes the Practice as Research PhD program at the University of California, Davis.

The final seven chapters are comprised of work in community performance, performed theory, performance art as research, performative ethnography, music and interactivity, PAR dramaturgy, and art/science collaborations. Marilyn Arsem writes about the way she conducts a collaborative embodied research with her audiences in order to explore and mark sites of erased/embedded political and cultural history; Shannon Rose Riley contributes a chapter on the use of performance art and performance as research in developing an ethnographic practice in Cuba; and Julie Wyman discusses her project, "Be Like Water," which conducts a set of "political investigations using speech in various forms (respeaking archival recordings and holding public meetings in charged locations [...])." Lynette Hunter, talks about her work in performing theory and community performance, and Jon D. Rossini's chapter discusses what it means to engage a performance as research approach to dramaturgy. The final two chapters in the collection approach PAR in relation to technological research. Joseph Butch Rovan describes his work in performance-based research on music, interactivity, and human gesture, and Sharon Marie Carnicke describes a collaboration between scientists and actors in a virtual reality, Star-Trek style "holodeck" actor training space.

Notes

1. *See* http://www.bristol.ac.uk/parip/. Also, see The Palatine Project at: http://www.palatine.ac.uk/directory/index.php/res/.
   In terms of music in the UK, *see* http://www.jisc.ac.uk/whatwedo/programmes/programme_rep_pres/repositories_sue/primo.aspx. The College Art Association in the US uses the term Studio Practice as Research. *See* http://www.collegeart.org/blog/2006/02/studio_practice_as_research.html. According to a Wikipedia entry in the UK, Screen Media Practice Research is defined as "an emerging research area situated primarily within Media Studies, Communications, Cultural Studies, Art and Design, Performing Arts departments in universities in the UK and around the world. It is difficult to define screen media practice research. At the most basic level it is academic research that is conducted in or through the practical production of film, video, Internet, visual arts and other screen-based media. It is a

subsection of a wider body of practice research within the arts and humanities." *See* http://en.wikipedia.org/wiki/Screen_media_practice_research.

2. For example, practice-based research networks (PBRNs) are groups of health care providers or medical clinics that typically practice in non-university based community environments. They aim to evaluate the health care processes that occur in real world practices. PBRNs provide a kind of "laboratory" for studying patients and care providers in community-based settings. The term "Art-Practice-based Research (APbR)" is also in use. *See* http://www.sunderland.ac.uk/~as0bgr/learnma2.html#quotes for some interesting quotes on this particular formation. *See also* http://www.ahrc.ac.uk/cpa/ for the language of practice-led research as framed by the AHRC in the UK. To further complicate things with an analysis of "research-based practice" in design, *see* http://64.233.167.104/search?q=cache:y3apULJgTZMJ:www.informedesign.umn.edu/_doc/Research_101_Part_I.pdf+%22practice-based+research%22&hl=en&ct=clnk&cd=15&gl=us&client=firefox-a.
3. Barbara Kirshenblatt-Gimblett, "Performance Studies (1999/2002)," in *The Performance Studies Reader*, ed. Henry Bial (London and New York: Routledge, 2007).
4. Ibid., 44.

Part I

Lay of the Landscape – Geographies, Legacies, Disciplines

1

Performance Practice as Research: Perspectives from a Small Island

Baz Kershaw

Preface: a playwright, a philosopher and a cycle ride

> At this time he and I were involved in a bicycle accident, which I feared for a moment might have brought his career to a premature end. He was only just learning to ride a bicycle, and he ran into my machine with such force that he was hurled through the air and landed on his back twenty-feet from the place of the collision. However, he got up completely unhurt and continued his ride. Whereas my bicycle was smashed, and I had to return by train. It was a very slow train, and at every station Shaw with his bicycle appeared on the platform, put his head into the carriage and jeered. I suspect he regarded the whole incident as proof of the virtues of vegetarianism.

The wry wit of that final judgment barely disguises a deep disdain reaching well beyond the virtuous cuisine of a man who 30 years later, in 1926, was to win a Nobel Prize. Twenty-four years later, within a month of G. B. Shaw's death, the philosopher also received the Swedish accolade. The biking reminiscence was written in the mid-1950s, 60 years after the crash that had the philosopher still troubled enough to ironically exaggerate the likely Shavian flight-path. Their age difference at the time no doubt added an anxious twist, with Shaw at 40 being 16 years older than his riding companion. But what to make of the pantomime that followed the accident, each station stop the scene of a repetitive window-framed put-down?

Obviously Shaw's performance had struck the great philosopher as something extraordinary, a weird event against which the best defense in retrospect boils down to a weak joke about diet. The youthful Lord Russell seems to have been rendered speechless by the older man's show, action-based effects prizing open his philosophical grip on the world, perhaps. But if that was the case, where might any knowledge produced by Shaw's action be located: was it in his resilient body, in the two minds of these remarkable men, in the story as told by Russell,[1] in this subsequent analysis – or hovering in the network of these options and more? That *dis-location* of knowledge becomes critical to its nature. The moment that Shaw's head enters the carriage his peripatetic platform performance becomes a prospective instance of practice as research.

Introduction

Such dislocation of knowledge by action is characteristic of performance practice as research, especially in its more radical forms. Russell's story reflects this, as it presents a specific space that is placed nowhere, neither rural nor suburban nor urban. As a result, any facts, truths, ideas, principles attributable to the scene become as fleet and wayward as autumn breezes; like all performance, there but not there. Hence both "original" actions and subsequent words produce undecidability in Derrida's sense: implying a decision that makes us responsible for their meaning.[2] This is why Derrida denied deconstruction was apolitical or a despairing creed, because such dislocations open the possibility of positive effects. He writes: "Any event brought about by a performative mark, any writing in the widest sense of the word, involves a *yes*."[3] Performance/performativity *as research* tends to be especially challenging to the academy in this regard, as its precipitate dislocation of established knowledge is positively foundational. No wonder that stiff resistance from academic traditionalists seems to have strengthened performance practice as research as an international phenomenon.[4]

The downside of this success, conventionally considered a democratic value, is interpretive freedom, a limitless general play of meaning. Often in the actual processes of creative practice this translates into profound unpredictability. Think of Harold Pinter's frighteningly rigorous "I don't know what kinds of characters my plays will have [...] I follow what I see on the paper."[5] These common vectors and factors, when combined as *principles of research* by serial performance practitioners, predicate endless differences and have spawned a fabulous diversity of practices. So performance practice as research partly aligns with performance studies and Richard Schechner's view that "there is no cultural or historical limit to what is or is not 'performance.' "[6] Performance is always already boundless. Yet also performance (and performativity) in both organic/inorganic "worlds" is never other than perfused by space and time. Every example is incorrigibly particular. Hence *boundless specificity* is a constitutive paradox of performance and performativity, creating multiple ontologies and epistemologies, ways of being and knowing.

That paradox ensures that performance practice as research – defined as the uses of practical creativity as reflexive enquiry into significant research concerns (usually conducted by "artist/scholars" in universities) – will present both highly specific and very broadly applicable results. Its concerns may be defined by the languages of particular disciplines, but the paradox also ensures this type of research will always resist becoming *a single discipline* as its projects proliferate insights, understandings, knowledges that, overall, will be part of many disciplines. Similar tendencies prompted Dwight Conquergood to propose performance studies as an *antidiscipline*.[7] A similar move for performance practice as research would claim it as a *transdiscipline*.[8] Implying that, most typically, it will generate procedures and protocols relevant to research in many disciplinary fields. Thus *performance practice **as** research* more precisely defines itself as method and methodology in search of results across disciplines: a collection of *transdisciplinary* research "tools." Boundless specificity produces precise methodological opportunities generally and a plethora of insights, understandings, knowings relevant to a wide range of disciplines specifically.

From this perspective, general criteria of value for performance practice as research results are a fool's illusion. But that does not make comparative judgment between specific projects impossible. Aspects of method should be broadly identifiable. Each project has to start in some way, and its enquiry will likely engage with aesthetics, location, affect/effect, documentation, dissemination, for example. This chapter briefly surveys some critical dynamics of such *aspects* through three examples of practice. Inevitably, given their brevity, these will be speculative reflections, meant to prompt further debate.

Questions of *methodology* – understood as systematized groups of methods exemplifying, or organized in terms of, particular theoretical perspectives – are much trickier to deal with. They might be clarified skeletally here, though, through the terms of reference informing this section: landscape, geographies, legacies, disciplines – which have guided my choice of examples. And there, of course, is where the methodological trickiness starts. For if, as I believe is the case, practice as research practitioners are like Gregory Bateson's explorer – "In the nature of the case, an explorer can never know what he is exploring until it has been explored,"[9] – how might one evolve a reflexive meta-account of methods arising from three unconnected, already completed explorations? The attempts of my analysis in this chapter in part constitute a "paradoxology of performance" that is homologous to the paradoxical nature of performance itself.[10] If boundless specificity is one of performance's crucial qualities, then a research methodology that plays on the insights created by paradoxes might match its complexities, especially when viewed from an ecological perspective that recognizes its full interactivity with "context." Then the contradictions raised by the dislocations of performance practice as research may become productive, especially if reproduced in the *structures* of an analysis. Hence, just as in gardens pruning creates growth, in performance the negative feedback of paradox may sustain its futures positively. In what follows, I hope, the irrelevance of that point may become blunter than it seems.

Starting points/location – Uninvited Guests and *It Is Like It Ought To Be: A Pastoral*

The small audience enters to be met by performers in traditional English rustic garb, who offer Elderflower Cordial or a cup of Ale.[11] The ground is green-grocers' grass, fake roses and vines twist up microphone stands, a stuffed horse drapes over an amplifier, autumn leaves and flower bouquets everywhere. They try apple bobbing in a metal basin before a long-skirted female shows the technique, long hair held back by a man in muddy boots and tweed trousers who suddenly violently pulls her head backwards then forces it down underwater as she thrashes about till drowned (Figure 1).

Ushered to the raked bank of seats, they become complicit witnesses. Astride the stuffed horse a performer stage-whispers into the ear of one that this "love horse" will pass on amorous messages; a tiny recorder captures her words, then is cantered away and played loudly to a third-row stranger. Recording animal sounds on a dozen machines, the cast conjures up a sonorous farmyard. Electronic hopping bunnies fill the level grass, serenaded by an old horn-violin. Climbing the "hill" of the seating bank, two locals

Figure 1 Uninvited Guests, Apple Bobbing Scene in *It Is Like It Ought To Be: A Pastoral*, directed by Paul Clarke, 2006. Photograph by Ben Pacey. Reproduced by permission of Paul Clarke and Ben Pacey.

warn of a huge storm brewing. The love horse is thrashed, its flesh becoming bunny food as dead leaves swirl in an electric fan's blast and a horse-headed demon goads semi-naked figures to fight like "wild animals."[12] Actions hang together as classic English "rural" in a randomly sequenced idyll-gone-wrong, deeply dislocated by the technically coy folkloric "staging."

The premiere for this touring show was a symposium at Lancaster University, England several years after director Paul Clarke secured a practice-based doctorate from Bristol University. A paper presented by Clarke at the event gives no clue about its specific starting point; its collaborative complexity suggests myriad possibilities.[13] Yet much simpler practice as research projects produce a similar problematic, with profound implications for research funding allocations before they start, as well as for doctoral assessments at their end. In the United Kingdom, the starting-point norm set by the Arts and Humanities Research Board (AHRB; founded 1998) was the same for PhDs and postdoctoral major grants: research questions. Yet the regular national RAE had been more liberal, in 1996 asking practitioner-researchers for a "succinct statement of research content" together with "supporting documentation,"[14] and in 2001 advising that practice as research should "interrogate itself" and be "driven by a research imperative."[15] But in 2008 it narrowed to asking for a "descriptive complement" relating to "questions addressed."[16] Meanwhile, the AHRB (awarded Council status in 2005) allowed researchers to identify "issues" and "problems" as well as "questions" in their applications.[17] Official systems of judgment shifted from a relatively liberal framework to one much less open to the inbuilt diversity of performance as research as actually practiced.

The paradoxes linking systems and practices here are posed, ironically, by the deliberate and contradictory mismatch between imagined location and the means of presenting it in Uninvited Guests' *Pastoral*.[18] To become a part of the scene the performing figures set themselves apart from it. By abstracting from it, the system shapes the field of research. Both system and figures reference an ideal in which stability arrives through predictability. If the weather "holds true," it will sustain the idyll. If questions are "answered" and problems "solved," then knowledge will be sound. But to "resolve" the paradox in these ways is an exercise in fantasy, because the actual locations, the environments of both figures and systems are inescapable. The differences between them hinge on their respective responses to this ecological fact, whether of "natural" or "cultural" environments. Thus institutional systems behave *as if* the research environment can be controlled – that is, by investing in it – while the figures on stage behave *as if* the research environment is beyond control, that is, by giving up to it. The underlying factors of process in this difference are profound: the first rests on vicious circularity (fiscal economies being ultimately uncontrollable) which produces a double bind and an addiction to performance as "valid results";[19] the second rests on a virtuous circularity (accepting the condition of uncontrollability) that produces paradox and creativity as a form of radical freedom. These formulations together suggest what may be at stake in the location of projects in performance practice as research. In this book's terminological framework, formally similar "landscapes" may generate deeply conflicting kinds of "geography."

Documentation/dissemination – Mike Pearson and *Bubbling Tom*

It is 25 April 2000 and Mike Pearson – dressed in a sharp buttoned-up suit, pressed white shirt and neatly tied tie – is standing in the middle of the Hibaldstow Beck, water up to the calves of his incongruous pair of Wellington boots, for all the world looking like a shaved-headed businessman caught in a practical joke.[20] He's saying, equally incongruously: *"Where to find pigeon's eggs, carrying them down from the nest in your mouth. Where to catch sticklebacks, the throats of the males blushed red in the breeding season. Or stone loaches camouflaged against the bottom of the stream."*[21] A small audience watches from the lane called, logically enough, Beck Side: his mother, wife, brother, aunt, uncle, cousins, old school friends, and sundry visitors from Sheffield and London. Only three photographs of this performance are in the book just quoted: Mike sitting in railings near the road junction to Scrawby; Mike gesturing up the street outside the Station Fish Bar; Mike, with fingers curved as if holding a football, close to the spring called Bubbling Tom. *"They say if you drink from Bubbling Tom you'll always come back. I don't think I ever did, probably because I was never quite sure where it was."*[22] The words of the show resonate strongly through the images, as if saying here's a canny man who really knows his place. But on page 54 there is a strange dislocation, because the photograph there is not of Mike standing in the Beck, but of someone else and its caption says: "Dee Heddon re-performing *Bubbling Tom*, Hibaldstow, 6 April 2002." Given the absence of Mike in the Beck, the Dee doubling is *uncanny*. She's Mike and not Mike.

In Comes I, the book that documents this practice as research, says *Bubbling Tom* was performed just twice. And, like Dee Heddon, I did not see either event. For believers in the theology of ephemerality, whose central tenet is that performance essentially does not survive its passing,[23] all Heddon or I can offer regarding this show is second-best goods, nothing of the original event. Even those with memories of it cannot retrieve its most crucial quality: "liveness." What kind of folly is it, then, to focus a *whole book* around a few performances? For Pearson the Professor of Performance Studies it is a major event, as sole-authored monographs possess top-dog research value in the humanities and social sciences. For practice as research doubters of documentary forms, though, it may be literally a sell-out. "If *only* artifact-based outcomes are seen to 'embody' the research as the 'serious' output, we might suggest that that reproduces the systems of commodity exchange."[24] Add influential theories about quotidian performativity stemming from Judith Butler,[25] and the "documentation" of live performance easily can seem tainted by technology and teleology. If the quality of a book provides any "proof" that *ephemeral* performance could produce *lasting* "knowledge" as an *ultimate* goal, then the devil of commodification always already would be dancing rings around the conflagration of radical hopes. From this perspective, just as the advent of the quill pen betrayed oral culture, any creative analogues between scriptural object and *Bubbling Tom* as live event will constitute a lost cause.

The paradoxes linking live practices and the nature of documents/dissemination here are generated by a further deliberate mismatch or contradiction, in this case between time/memory and space/embodiment. The performing figure becomes integral to the scene by being abstracted from it. The document/dissemination manifests the "live" precisely through its absence. The book achieves this remarkable feat through a double movement, whereby the more Pearson evokes the past, the less "he" is directly represented (by description or images) *and* the more he is present as "author." Space/embodiment is implied through a near-obsessive documentation of location whose methods function as a paradoxical reversed telescope effect. Each of the book's three sections increasingly pull focus on the minutia of memories in *Bubbling Tom* to create a widening panorama of landscapes and disciplines, so the performing figure/Pearson, as it were, shrinks to the tiniest of specks in expanding space. Likewise, the focus of time/memory expands from Pearson recalling childhood events to accounts of prehistoric scenes, encompassing archaeology, geography, geology, natural history, folklore, antiquarianism, and more in its transdisciplinary compass. Doubters of the document as reactivator of living memory might see failure in the fading figure of the performer, but they overlook the paradoxical power that challenges the law of the excluded middle, that no proposition can be both true and not true. Hence just as a person *may* see the whole of their life flash past in the moment of death and thence enter a living eternity (as the memory sequence must include itself), so the disappearing figure of performance may be magnified beyond imagination into life. *Bubbling Tom* thus *gains* power because it is rendered uncanny, always there-and-not-there in the vista of Pearson's recreated "landscapes." The minimalism of pictures and words becomes phenomenally charged with the living energies of their past event. Hence, in this section's terminological framework, "geographies" might mightily reactivate "legacies" even through the faintest traces of performance.

A test conclusion – Baz Kershaw/Sandra Reeve and *Being in Between*[26]

The Czech spider monkey was in quarantine, alone in a large glass enclosure, with ropes and ladders and vegetation. In front of the enclosure was

> a large square of open ground. The performers worked to establish communication with the monkey through non-stylized movement, initially in the mode of primatologist.
>
> In the final performance there were around fifteen people consciously there as audience. Oblivious visitors walked into the zoo without noticing the dancers and went to the glass front of the enclosure to look at the spider monkey. When they became aware of the dancers, either through the spider monkey's activities, the reflections of the dancers in the glass or the suspended atmosphere of incredulity, they found they had placed themselves in a network of interactions. Reactions to this ranged from slight indignation or embarrassment to an interest in what was happening. A little boy stood up and joined the performers, imitating their movements. He was consciously included in the dance. Front of house staff slipped out to watch and zookeepers appeared – word had got out that there was a magic moment when the performers and spider monkey danced together. As people perceived the undisputable *relationship* between the humans and animals through movement, and witnessed the change of role as the humans danced for the monkey, the dance began to make sense in a different way and comments changed from "oh look they are pretending to be monkeys" to "oh look they are dancing with the monkey!" In those moments, awareness and meaning shifted [...] I could suddenly hear the sound of the wind in the leaves of the large beech tree and see the expression of delighted amazement on my father's face.[27]

I conclude with this scene as providing a skeptical test of my "paradoxology of performance" methodology. Paradoxically, the scene's effects invite skepticism *precisely because* its aesthetic success depended on the performers' literal disappearance for the zoo's visitors as they performed. This was mysterious because costume and movement made them highly visible in their difference, yet *always* there were some visitors who were "oblivious," who initially seemed simply *not to see them* even though they might pass very close to the dancing (Figure 2). As if some quality in the monkey's movements must be mesmerizing them. Speculation suggests, somewhat fancifully, that this was an echo of human movement, as the monkey usually was obviously responding to the dancers. So were these visitors drawn by animal magnetism into some paradox of the human in the non-human? Had the monkey's performance somehow caused the *Being in Between* human performers to be there-and-not-there, making them phantom performing primates?

Of course, such speculation is ridiculous as, by definition, if it is correct – *especially* if correct – there is no way it can be verified by those who were mesmerized. So any "knowledge" arising from that profound dislocation through performance as practice as research cannot be confirmed:

Figure 2 *Being in Between*, co-directed by Baz Kershaw and Sandra Reeve, 2005. Photograph by Drew Yapp. Reproduced by permission of the photographer

one can only indicate the conditions that may have made it possible. But nothing in the conditions will automatically dispel skepticism. End of story. Yet if there is even the slightest of chances that, given appropriate conditions, paradoxes can produce sound truths, as the philosophical approach called "dialetheism" argues,[28] then the paradox of boundless specificity that characterizes performance – which in practice as research creates dislocated knowledge – may sometimes meet those conditions. Paradox: the future is always before its time. Conclusion: through performance all primates evolve human qualities. In this ecology animal "legacies" may well become human "disciplines."

Postscript

Philosophically speaking, it will never be known for sure that Bertrand Russell's knowledge was unfixed by Shaw's ridiculous accident-inspired jape. But Russell also wrote that the latter "was at his best as a controversialist."[29] And a few years later he said of himself as a young man: "I had passion and fed on controversy."[30] Perhaps it was, after all, Shaw's unsettling gift of performance that in part spurred on Russell's famous lifelong passion in the search for knowledge.[31]

Notes

1. Bertrand Russell, *Portraits from Memory* (London: Readers Union, George Allen & Unwin, 1956), 71.
2. Jacques Derrida, *Acts of Literature*, ed. Derek Attridge (New York and London: Routledge, 1992), 172–5.
3. Ibid., 298
4. Ludivine Allegue et al., eds, *Practice-as-Research: In Performance and Screen* (Basingstoke: Palgrave Macmillan, 2009).
5. Ian Smith, ed. *Pinter in the Theatre* (London: Nick Hern Books, 2005), 52.
6. Richard Schechner, *Performance Studies: An Introduction*, 2nd edn (New York and London: Routledge, 2006), 2.
7. Dwight Conquergood, "The Institutional Future of the Field," at *First Annual Performance Studies Conference: The Future of the Field* (New York: Tisch School of the Arts, 1995); *see also* Joseph Roach, "Culture and Performance in the Circum-Atlantic World," in *Performativity and Performance*, ed. Andrew Parker and Eve Kosofsky Sedgewick (New York and London: Routledge, 1995), 46; Nigel Thrift and J.-D. Dewsbury, "Dead Geographies – and How to Make Them Live," *Environment and Planning D: Society and Space* 18, no. 4 (2000): 420.
8. Basarab Nicolescu, *Manifesto of Transdisciplinarity*, ed. David Appelbaum, trans. Karen-Claire Voss (New York: SUNY Press, 2002).
9. Gregory Bateson, *Steps to an Ecology of Mind* (Chicago: University of Chicago Press, 2000), xxiv.
10. *See* Baz Kershaw, "Performance Studies and Po-Chang's Ox: Steps to a Paradoxology of Performance," *New Theatre Quarterly* XXII, no. 1 (2006); and his *Theatre Ecology: Environments and Performance Events* (Cambridge: Cambridge University Press, 2007).
11. *It Is Like It Ought To Be: A Pastoral*, Uninvited Guests Theatre Company, director Paul Clarke, premiere performance 4 September 2006, Nuffield Theatre, Lancaster University, then touring 2006–8.
12. Eleanor Hadley Kershaw, "Uninvited Guests: *It Is Like It Ought to Be: A Pastoral*," (Arnolfini, Bristol, 2006). Unpublished review.
13. Paul Clarke, "Between You and Us: A Story of Uninvited Guests," Uninvited Guests, http://www.uninvited-guests.net/Symposium/PaulClarke.html.
14. RAE, "Criteria for Assessment 66 Drama, Dance and Performing Arts," HERO, http://www.hero.ac.uk/rae/rae96/66.html.
15. Ibid., "Section III: Panels' Criteria and Working Methods," HERO, http://www.hero.ac.uk/rae/Pubs/5_99/ByUoa/crit66.htm.
16. Ibid., "UOA 65, Drama, Dance and Performing Arts," RAE, http://www.rae.ac.uk/pubs/2006/01/docs/o65.pdf.
17. AHRC, "Research Grants," AHRC, http://www.ahrc.ac.uk/apply/research/research_grants.asp.
18. *See* www.uninvited-guests.net
19. Kershaw, *Theatre Ecology*, 13–5.
20. *Bubbling Tom*, writer/performer Mike Pearson, performances 24–25 April 2000, Hibaldstow, Lincolnshire, UK.
21. Mike Pearson, *In Comes I: Performance, Memory and Landscape* (Exeter: University of Exeter Press, 2006), 55.
22. Ibid., 71.
23. Peggy Phelan, *Unmarked: The Politics of Performance* (New York and London: Routledge, 1993), 14.

24. Angela Piccini and Baz Kershaw, "Practice as Research in Performance: From Epistemology to Evaluation," *The Journal of Media Practice* 4, no. 2 (2003): 122.
25. *See* Judith Butler, *Gender Trouble: Feminism and the Subversion of Identity* (New York and London: Routledge, 1990); Andrew Parker and Eve Kosofsky Sedgewick, eds, *Performativity and Performance* (New York and London: Routledge, 1995).
26. *Being In Between,* co-directors/devisers Baz Kershaw and Sandra Reeve, performances 14–16 October 2005, Bristol Zoological Gardens, Bristol, UK.
27. Sandra Reeve, "The Next Step: Eco-Somatics and Performance," *The Changing Body seminar* (Exeter: University of Exeter, 2006), 18.
28. Graham Priest, *In Contradiction: A Study of the Transconsistent* (Dordrecht: Nijhoff, 1987); Kershaw, "Ox."
29. Russell, *Portraits*, 74.
30. Ved Mehta, *Fly and the Fly-Bottle* (Harmondsworth: Pelican Books, 1965), 44.
31. Bertrand Russell, *The Autobiography of Bertrand Russell: The Early Years, 1872–World War I* (New York: Bantam Books, 1968), 3.

2
Working on the Middle Ground: A Case Study of Institutional Inter-Action about Practice as Research

Simon Jones

Along with much writing alongside *(my preposition of choice) practice as research [PaR], I have tended to the extremes: either the dizzying heights of theorizing,*[1] *or the nitty-gritty self-reflexive explicating of particular practices, especially my own with Bodies in Flight.*[2] *This verticality of writing, as it were from the two ends of the scale – macro–micro – inevitably leaps over the middle. This is odd given the heightened significance of the necessary pragmatics of how one actually gets to the point of making a work, of becoming a practitioner-researcher. So, at the risk of being mundane, I want to write a short story, or rather – let others now involved in the same story speak it on all our behalves, about the middle ground of PaR: the horizontal plane or "ground level"*[3] *of funding and networks, places and persons, opportunities and risks. Without theorizing too much, I want to make an account of the de-Certeaudian tactics of the non-teleological, Deleuzian rhizomatic assemblage that has come to happen between the University of Bristol's Department of Drama: theatre, film, television, and Arnolfini: the one being the first university drama department in the UK* **[established 1947] up on the hill** *and the other being a leading contemporary visual and performing arts venue* **[established 1961] down** *by the harborside of this small West-Country city.*[4]

And so, as the critical mass of PaR is building **here** *both in terms of the macro-level strategizing between the two major funders on either side of the academic–professional divide (the Arts Council of England [ACE] and the Arts and Humanities Research Council [AHRC]) and in terms of the micro-level gatherings of practitioner-researchers, curator-producers, archivists, students, and public audiences, it seems most apt at this point, whilst acknowledging my own role in forging these links and developing these relationships, to populate this narrative with the voices of some of those who are following on and making this community through the depth of their ambitions and the reach of their imaginations.*

My hope is that this story of the mix of happenchances that catalyzed individuals, projects that synergized efforts and PaR artworks that galvanized communities of practice will inspire those of you who do not currently work in such a benign environment to persevere, have the courage of living and working in-between the various flows and forces of capital, to forge the personal connections, seize the funding and showing opportunities, and make the new kind of community that practitioner-researchers need in order to thrive.[5]

HELEN COLE (*Dance and Live-art Producer, Arnolfini*): The advantages of working with the Drama Department are many and multi-various. They range from the subtle interventions that happen that actually affect the community of practice, the artists we engage with locally and internationally, to the very early points of working with students who are finding their first professional platform for making work. That is, the intrepid, entrepreneurial student who trips **down** the hill and stumbles upon the doorway of Arnolfini. For instance, Paul Clarke, who forwarded a piece of documentation as well as coming to see work.

PAUL CLARKE (*Research Fellow of Performing the Archive at the Department of Drama and director of Uninvited Guests performance company*): I started a PhD **here in** Bristol entitled "Collaborative Performance Systems," and it was about a year in that the relationship began between Uninvited Guests and Arnolfini. My PhD included practice that I wrote around, rather than being formally included. At that point **[1998]** it was not possible to do a PaR doctoral project at Bristol. I was interested in setting up my writing within the PhD and my practice as two parallel strands of practical methodologies, rather than attempting to illustrate the theory through practice, or equally to analyze the practice in the theoretical writing. However, I was uncomfortable about how the practice was included in the PhD as reflections within the writing and documentation of scripts, videos, a CD-Rom as appendices. Now I would make that more central and name that as one of the research methodologies that would be part of, rather than an appendix to, the thesis.

HELEN: When I arrived in Bristol **[1998]** I compared the context **here** with what I had known in other cities – Glasgow, Manchester specifically. Interestingly both do have a variety of practices that emerge from the academy; but the relationship between ourselves and the Department is more particular, has more breadth to it. That started quickly through those very organic connections between students, doctoral researchers, us engaging on major projects together [***The Goat Island Summer School: Bristol 2000–2*** *(funded by ACE, Esmee Fairburn Foundation and the AHRC Research Exchange Scheme);* ***Performing the Archive: The future of the past*** *(in partnership with Arnolfini, Bristol and Exeter Universities, funded by Great Western Research, itself funded by the Higher Education Funding Council for England and the South-West Regional Development Agency)*].

BEX CARRINGTON (*Keeper of the Live Art Archives [LAA] held at The University of Bristol Theatre Collection, a museum-status archive located within the Department*): We have a student rep on the Theatre Collection Management

Committee who is very keen on the LAA, including tours with the freshers. This helps the confidence of other students to explore the holdings. They then go **down to** Arnolfini to see performances; and I send students off to the Live Art Development Agency's [a London-based agency whose mission is to foster live-art nationally] extensive reading resource, and they then come back and ask to see specific performances held in the LAA. This works well. By all means, I treat the three organizations as a whole resource. We are all trying to get the information out there to people as best we can. There's no competition because the core functions are different and complement each other.

JO ELSWORTH (*Director of the Theatre Collection*): The LAA have also attracted interns, themselves performing-arts graduates, to volunteer in the Collection, working on a regular basis on particular projects. This is new and exciting. For instance, one is scanning *Performance Magazine* in preparation for on-line dissemination, reading articles as he goes. He is also reboxing the records of LA practice. The other intern is now working as part of the project-team ***Capturing the Past, Preserving the Future: Digitizing the National Review of Live Art Video Archive*** [*an AHRC-funded resource-enhancement project hosted by the Department, the purpose of which is to preserve the video records of this major performance festival by migrating them from analogue-tape formats to a digital platform*].

HELEN: In terms of joint projects, for instance, with Goat Island, we were showing the work, but because they are also researchers and teachers, there was another facet to their work that we could not really engage with to the depth that we eventually did when we started working the Department. The first symposium we held as part of the summer school **[2000]** was really important because it brought together practitioners with scholars, including national and international participants. And I first encountered other artists through those events, for example – Lone Twin presenting a paper as practitioners in the symposium. In performance studies, there is not such a clear boundary between practitioner and scholar; it's very fluid. So, I was first introduced to Lone Twin within that academic context, through Goat Island's relationship with David Williams at Dartington College of Arts. Indeed, the summer schools affected a whole generation of artists. And the longer-term collaboration between the two institutions, since hosting its first Goat Island summer school **[1992]**, has meant that it has had an impact on several generations of artists. I remember Mark Jeffries [*a company member who first met Goat Island* ***then***] speaking to me about how important that moment was to him. There has been influence from both sides [*The beginnings of the relationship between Arnolfini and the Department* **(1990–7)** *had foregrounded professional practice, as in co-presentations at the Department's Wickham Theatre of artists, such as DV8 and Goat Island, and the commissioning of Bodies in Flight's work, most recently* Model Love **(2008)**.]

When I compare my experience as an audience-member at other venues, the level of critical engagement **here** is much higher. I think that's because

Figure 3 Bodies in Flight, *Model Love*, Arnolfini, 2008. Photograph by Edward Dimsdale

of the thinking happening here. We get infected by each other. I notice this more in the artists I work with, the way they talk about their work, the other artists they line themselves up with, rather than with other producers, even though they may themselves be operating in an academic environment [*a venue sited and/or funded within a university context: for example, in the UK, Lancaster's Nuffield Studio or Manchester Metropolitan's Alsager Arts*]. For these artists, it is simply not just coming to see a show: a big part of the experience is talking about the work in the bar immediately afterwards.

JULIAN WARREN (*has worked at Arnolfini since the mid-1990s and in 2007 was appointed its first archivist*): It is also because the dialogue is sustained. There's a shared knowledge going on. We return and keep critically engaging with work over a period of time. With this community of artists and people from the university. It helps that Bristol is not such a big city, and that the university is just **up the hill**.

SHI KER (*a PhD student whose PaR project explores absence and presence across a range of performance, live-art, and virtual environments*): I find it not so revolutionary. If you look at science or engineering, then they have been doing these collaborative schemes for a long time. You have to produce something to prove your theoretical thinking.

HELEN: To have one of the leading performance studies departments in the country and a leading professional platform spawns an ***associate-artist***

program with 18 artists and growing. Most small cities would not have that number. There has been an interesting dialogue between the academic institutions in the region and Arnolfini [*University of Bristol, Dartington College of Arts, University of West of England*]. It has been a rich environment. People do recognize work that emerges from Arnolfini as having a particular character. We have made a contribution to British practice. It's unusual for any program to encompass the range of artists we do, for instance, body-based solo-artists like Paul Hurley [*a PhD on the* ***Curating Risk*** *project, funded by the AHRC as part of its Collaborative Doctoral Scheme fostering relationships between higher education institutions and the creative and cultural industries*] and Kira O'Reilly, alongside a theatre company like Bodies in Flight or Uninvited Guests. As well as the international artists, such as Ron Athey or Goat Island.

PAUL CLARKE: The Bristol performance scene was more welcoming than the London scene; there seemed to be more ways into dialogues with other artists who were excited by these conversations and the inspiration that they might produce. That was focused around both the Department with its graduates and Arnolfini with its program. It provided a social focus for those conversations around practice, especially when arriving in a new city.

JULIAN: And a lot of those collaborations start **here in** the bar. You sit with the audience after a show, members of staff, artists in the audience, academics, and academics who are artists, and everyone has a different approach.

BEX: Julian is much more high profile than myself, because he's **down at** Arnolfini, at events. He has different connections from myself.

JULIAN: Collaborations and friendships grow; conversations happen, you have drinks, talk about the work you've just seen. Unlike similar conversations in bars at other places, we see the possibilities and opportunities. That fluidity starts with someone like you, Simon, as practitioner and academic. The first show that I saw at Arnolfini was Bodies in Flight – *Littluns Wake* – the performer packed in ice. At that point **[1995]**, recently graduated and having seen a lot of theatre, coming into that situation it was obvious that this was a different, very critically engaged kind of theatre. It made me want to go and see everything at Arnolfini, because it was a theatre engaging with its means of production. Having studied English at university, focusing on play-texts, this was a revelation: a kind of theatre that only happens in a place like this, but comes out of a university with practitioner-academics like yourself.

PAUL CLARKE: As a young company we were surprised at how welcoming Arnolfini was, it being a venue that within the context of visual arts tends to show international work. That was Helen Cole's decision to open up the program to emerging artists, especially those of us working within the city and southwest region. Having sent her some documentation of *Detective Stories* [*at that time Paul's company were called Parasite*] and having seen an early work-in-progress of *Guest House*, she invited us in and asked what Arnolfini could do for local practitioners. This became ***Breathing Space*** [*a strand of Arnolfini's programming focusing on emerging artists*]; and we were the first group to benefit from a two-week residency at Arnolfini with an opportunity to share our work, framed as work-in-progress. For us, this was very helpful.

There was a real proximity between the way we were working at Arnolfini and what would now be called PaR. We were clearly being invited to use the space as a laboratory, to try things out that were conceptual in the way that they were framed, coming from a position where we did not know what the outcome would be. Partly because of Bristol's supportive community of practitioners, audiences would understand and get excited about coming into work in its early stages. They would respond critically through post-show discussions or feedback in the bar in such a way that they felt their words might actually have some impact. For instance, the durational version of *Offline*[6] emerged out of conversations with audiences as well as Helen as producer: it encouraged us to push that further, so that having been interested in what happens when you speak in a performance-event text written for online consumption, we got interested in what happens when you remove the theatrical encounter and allow the audience to come and go and speak with them much more intimately. Arnolfini offered us a space where we were comfortable with taking that risk. It was high profile, as we showed at Inbetween Time, an international live-art festival, but in a place that understood risk and experimentation.

HELEN: When I first arrived one of my key objectives was to try to bring more local artists into the program and create that community of local practice. So, I diverted a significant proportion of the programming budget to enable that. A good number of those artists were from the Department – Bodies in Flight, Uninvited Guests, Special Guests [*a performance group who graduated* 2003]. We have the Collaborative Doctoral scheme that encouraged more artists to move to Bristol – Paul Hurley and Steve Robins. Also Shi Ker's involvement with that scheme, although not funded through the AHRC. And now Associate Artists, such as Michael Jones, are going into the Department as research students [*completing a MPhil exploring intimacy in performance*]. It comes about through excitement and learning around a program of work, as well as a study of work.

PAUL CLARKE: I think there's a lot of potential for postgraduate students to work between the Department and Arnolfini, who might be drawing on the resources of the archives **here** and **there**, and also using Arnolfini spaces as a professional context for showing works-in-progress. It answers a concern that PaR can only function within the academy and does not speak professionally beyond that context to a wider public.

PAUL HURLEY (*whose doctoral project explores Deleuzian ideas of becoming through a series of performance works and self-reflexive writing*): Presenting my

Figure 4 Uninvited Guests, *Offline*, Arnolfini, 2001. Photograph by Tom Hall

work publicly at Arnolfini as part of the PhD and the feedback that follows is really valuable, because I am not just doing a private performance to fellow students. We are actually putting something out there, that primes us much better to carry on working. It also increases our public profile as artists.

SHI KER: It is this opportunity to experiment and to see what comes back, the feedback and reflection. It touches the real world.

RUTH HOLDSWORTH (*whose project explores notions of risk and curating live-art across various festivals and international contexts*): It was an experiment. You had no models to follow. It was part of the first tranche of funded collaborative PhDs.

PAUL HURLEY: There's the strategizing of the two institutions in putting the scheme in place [2005–9]; and then there's the individuals involved and the trajectories of each person's doctoral project. For me, I was wanting a freedom despite both institutions. Actually, my project needed that structure to break away from. And having a different relationship with each institution is also very useful. We are somewhere in-between the professional world and the purely academic. I did my masters at an art college, focused on professional practice at both undergraduate and postgraduate levels. Although there were not opportunities to show work professionally, there was that institutional mindset. PaR provides this space that is no longer given by the industry; and that's partly because of money, the way arts organizations are tied to their funders.

RUTH: And in terms of freedom, there's the way that time is structured within both institutions. Within the Department, taking your time to think is really valued. You might have one thing that you are seeding for three years. Within Arnolfini, despite the long timeframes for planning, there is a different sense of how you might ruminate about a project, how a project is realized. When they have an idea, it's realized in a public forum; whereas in the university the idea is realized in quite an insular, contained forum. So, there's different ways of approaching thinking and forming ideas. I have found that dual rhythm problematic, and have tried to compensate for it in my own professional career.

PAUL HURLEY: As well as the public aspects of what we do, there's also the conversations about my research and work, Helen Cole's understanding of it, which is different from Simon's. Those benefits are valuable despite being quieter, often informal and intangible. Having a relationship based on a research-led practice within an arts organization is unusual, because one normally just has a relationship about the work.

RUTH: The scheme allows you to fast-forward making your work public. You have a sense of the kind of connections and relations you hold. For me, the opportunity was the Lebanese artists event I organized **[2006]**. I had the idea and one week later it was happening. This allowed me to connect all the things I was thinking about, not reflect too much, and throw it out there to a public and see what came back from that forum. Being able to surface and say this is what I am doing during the process of doing it. On a conventional doctoral project that would probably come at a later stage. **Here** you surface publicly more frequently and then you have the time to take that experience and feed it back into what you are doing. For me that has been confusing at times, because I am constantly dealing with others' current work, I get so much back.

STEVE ROBINS (*whose project explores ideas of beauty and justice in Live Art through both practice and theoretical writing*): *Niceday* [*a performance with Shi Ker, part of both theses*] was launched into a different realm, after a work-in-progress showing at the Department, and then being shown at Arnolfini **[2007]**.

SHI KER: Yes, without Arnolfini, that would have been a full stop, without the opportunity for feedback, to make a further move and improve it. As a Chinese artist, you have to foreground your Chinese identity to establish yourself in the Western art-world. But with Arnolfini I got this chance without using my Chinese identity, but through PaR directly into an

Figure 5 Shi Ker and Steve Robins, *Niceday*, Arnolfini, 2007. Photograph by Carl Newland. Reproduced by permission of Arnolfini

art-space. So, it provided a greenroom, without being categorized as either artist or intellectual.

JULIAN: With the interrelationship between the archives involved in the Performing the Archive project, there is a growing critical mass. Arnolfini doesn't have what you might call a proper archive yet; and my post is new. Arnolfini has a collection of things. I have to figure out how we turn this collection into an archive. The conversations I have with Paul Clarke, Bex and Jo, Barry Parsons and Stephen Gray [*Research Associate and Conservation Technician on the Capturing the Past project*], all help shape and inform the ways in which we might want to take things forward. There is a real productive exchange. I can say to Paul – well, no, pragmatically that's never going to happen. But he can still go ahead and explore it theoretically, and introduce things that I had not thought of, and make me rethink. It's really exciting. It pushes our archival practice. And in Arnolfini as an organization that is always looking forward rather than backwards, it's good to have that critical partner and help discover ways to approach the archive.

JO: There are no tensions between what we are trying to do with the archives in a university, research-oriented context, and what Julian is trying to do within an arts organization.

HELEN: We have benefited from the visual arts greater respect for objects. So, Arnolfini has always collected documents and artifacts from its activities, making it unique as it crosses the various disciplines. And the association with the university helps sustain that.

BEX: From my point of view, since moving with the LAA from Nottingham Trent University to Bristol **[2005]**, the relationship has grown so much, not only from the point of view of liaising with Julian, but also being invited along to events, either research-based or otherwise, where you meet a network of people who connect back to other things like the Now Festival at Nottingham [*a leading annual festival of contemporary dance and performance*]. It has shown me how rich our network can be. **Here** in Bristol we are creating a real hub now. A much closer relationship. The Live Art Archives were originally set up **[1994]** to record events; now it's a lot more than that. As well as feeling like you are part of something bigger within the Department with the research events and student work, we are working with Arnolfini to help create their archive, looking at what they have, what we have. For instance, we exchange the information content held in our archives with Arnolfini, such as a photocopy of a program or duplicates of printed material, if Julian finds they are missing material that relates more to their work and less to our own. From the LAA point of view, this exchange adds more context around the holdings; and from Arnolfini's it ensures that they aren't holding more than they need to. It is not a competition but a sharing of resources, something we are also cultivating with LADA, and also recently with Locus+, a Newcastle-based archive, which has just received Alistair MacLennan's archive.

JO: What we notice working with Arnolfini is that their archive is very integrated with current practice and practitioners which is unusual with archives. Even with our own archives of active companies [*for example, London Old Vic, Welfare State International*], we don't have that same link. This means that the tripartite arrangement, us as an archive, Arnolfini as the arts organization, and the Department's research focus, has created a real sense that the archive is dynamic and responding to and influencing current practice. We have a similar relationship with the Bristol Old Vic [*a local repertory theatre*]; however, it is not as responsive. Although their archive is being used for fund-raising, publicity, even investigations to guide the restoration works now underway at the Bristol Theatre Royal, it's not being used in a creative sense, but a very pragmatic one. They are foregrounding their history. However, because of the links with the creative-industries, the LAA tend to get used as a creative reference point.

BEX: When we get practitioners **up from** Arnolfini, then they see the rest of the holdings. The traditional function of the archive is to look back, whereas I see live-art practitioners and audiences coming in and seeing it as a living thing, a springboard for future work. Franko B was presenting at Arnolfini as part of the Performing the Archive project. After the work was shown and he had given his talk, he donated his archive to us.

JO: One of the reasons Franko chose to do that are those links with Arnolfini combined with the credibility we have as a research-led organization.

JULIAN: For the archive to have meaning it has to be reused and revisited, critically engaged with and so brought to life. The relationship with the Department enables those things to actually come about. For me, it's all about how we are going to push the archive in the future to push people to think about history and performance in new ways. In order to conserve the material, maybe the traditional focus of archives, the relationship with researchers helps equip me for developing the archive. For instance, the LAA, although they fit into a previously existing Theatre Collection, the way they are approaching conservation, which is incredibly novel and at the forefront of archiving [*the Capturing the Past project*], informs my work. One problem, it seems to me, is that the British Standards Agency standards are not yet adequate for archiving such artifacts as digital-media platforms. The digitization project is informing the development of those standards, impacting on Arnolfini's archive and so in turn on Bristol Records Office where the archive is held. But also we've been finding that it's often not the documents of the performances themselves that are the most informative, but the documents of preparatory materials, texts, treatments, that are being used the most. And this informs what we might document in the future.

PAUL CLARKE: What attracted me to the Performing the Archive research fellowship, having been working as a lecturer at Dartington College of Arts, was the relationship with Arnolfini and Uninvited Guests' interest in performance as archive or a performed archive. Many of our performances have engaged with re-enactment and also with the collecting of others' memories. Furthermore, the fellowship was interested in exploring the interconnections between professional practice and the academy, particularly with an arts organization interested in the arts as a mode of research. The structure of the archive is being influenced by research, particularly in its online presence, through the ACE-funded Thrive project that links the various arts organizations based around Bristol's harborside [*Arnolfini, Architecture Centre, Spike Island Galleries, Watershed Media Centre*]. The structure is being invented as we go along, responding to the needs of artists, researchers, and users.

HELEN: When I think of the outputs from Arnolfini, they are really diverse. They could be a show, equally a publication, an exhibition, a talk event, something in the archive. And that is much more similar to the way the academy works than a producing repertory theatre. That's because we work in an interdisciplinary way and deal with multiple audiences. We are influenced by other artforms and engaged in critical dialogues about our practices. I don't think there's clear blue water between what Arnolfini and the various practitioners and researchers in the Department want to do with the relationship. There are numerous bridges across. The university is a vital part of the public whom we are engaging. For me, the sharing of ideas about moving forward the discipline is really important. The political nature of the relationship is also important. Especially working in a performance discipline within a visual-arts led organization, it's hugely desirable that we have this shared territory with external partners who bring their knowledge and expertise into this place. Having a large archive **up the road**, and **here** an archivist incredibly knowledgeable about performance, strengthens the argument for the live-art program. It's about building a body of proof.

RUTH: The interesting thing is when you factor in personality.

HELEN: Amongst the top relationships I've developed since coming to Arnolfini, the one with the Department is ultimately about the individuals involved. It is with you. It's fragile because it's about individuals who share an understanding.

SHI KER: One of my favorite Chinese philosophers says knowledge and action are one and the same.

Notes

1. Simon Jones, "The Courage of Complementarity," in *Practice as Research in Performance and Screen*, ed. Ludivine Fuschini et al. (Basingstoke: Palgrave Macmillan, 2009).
2. *See* Sara Giddens and Simon Jones, "De-Second Naturing: Word Unbecoming Flesh in the Work of Bodies in Flight," in *Sensualities/Textualities and Technologies: Writings of the Body in 21st Century Performance*, ed. Sue Broadhurst and Josephine Machon (Basingstoke: Palgrave Macmillan, forthcoming); Simon Jones, "Imag[in]ing the Void," http://presence.stanford.edu:3455/Collaboratory/1173; ibid., "Places Inbetween: I Do Not Have to Be There to Be There with You Tonight, a Case Study of Bodies in Flight's Performance Skinworks," in *Collision: Interarts Practice and Research*, ed. David Cecchetto et al. (Newcastle upon Tyne: Cambridge Scholars Publishing, 2008); Josephine Machon, *(Syn)Aesthetics? Towards a Definition of Visceral Performance* (Basingstoke: Palgrave Macmillan, forthcoming); Sara Giddens and Simon Jones, eds, *Flesh & Text: A Document by Bodies in Flight* (Nottingham: Future Factory, 2001). *See also* www.bodiesinflight.co.uk.
3. Dwight Conquergood, "Performance Studies: Interventions and Radical Research," *TDR: The Drama Review* 46, no. 2 (2002).
4. *See* www.arnolfini.org.uk.
5. Interviews conducted in various locations in Bristol, October 2008.
6. The durational version of *Offline* occurred in 2003 – the show version in 2001.

3
Troubling Performance: Local, National and International

Richard Gough

I want to interlace this exploration of performance and research with three short stories based on real experiences of entering and exiting the United States, a series of encounters with immigration officers and homeland security guards that continue to excite, amuse, and terrify me.[1]

The value of place

Story One. Entering Atlanta, Georgia, 1997, for the Performance Studies Conference hosted by Georgia Tech. I seem to have been in a queue for hours. When I finally get to the yellow line – that I should not cross until I am invited to do so – and then do so, the official puts a red line straight through Wales – my country of origin – and over-writes *England*. I say, "You've just cancelled my country." He doesn't even look up, and sharply intones, "If you want to argue about it, go to the end of the line."

Restless from jetlag, half-dreaming, fitful and fleeting, in a cold sweat, I seemed to be on a border between theatre and performance, stopped once again at the security gate. Barefoot, belt-less, pockets emptied, laptop exposed, jacket removed, clutching simply a passport and a boarding pass (to where?), was I trying to enter performance from theatre or re-enter theatre from performance? Was Theatre about to have a line put through it with Performance overwritten? I pondered this palimpsest. It was at this moment the voice said "what do you want, what is your root, what is your destiny?" And did he mean root or route and did he say destiny or destination? It was the immigration officer from Atlanta again and I could sense the shrill order to return to the back of the line about to be barked.

I am from Wales; England's First Colony. My approach to performance is wrapped up in this fact.

In our work at the Centre for Performance Research (CPR), we take as our starting point the position of Wales on the periphery of Europe, and transform this into a curatorial vision as we take a broad look at contemporary performance work. We produce work which proposes new developments in artforms and between disciplines, and which reaches out to, and engages with, new audiences: dance/theatre; installation/performance; music and multimedia projects; site-specific and landscape-based events; work which places different body shape, age, color, and "ability" at its centre. Performance which makes the marginal central, celebrating diversity and all that which exists on the periphery, on the edge, on the border between different artforms and between social and aesthetic action – that which disturbs, illuminates, challenges the norm, takes a paradoxical position, is made off-centre, off-side, on purpose. Works which might previously have been thought of as disenfranchised, not just work which is "the stuff" of the international festival circuit, but work which is made in and from a particular and specific set of circumstances. Work which is distinguished by its own sense of displacement, which is angry and passionate, flagrant and partisan, speaking directly, visually, and viscerally to a wide audience.

CPR's roots are in a Grotowski/Barba tradition of physical experimental work which seeks the connections between innovation and tradition, and is curious to learn from the great world theatre traditions. It is keen to explore the potential of the open symbolic space of the stage, where the actions of the performer are central and the verbal dramatic text less essential; indeed, where conventional dramatic structure is displaced and an emphasis on what is essentially theatrical (as opposed to literary) is returned and reconstituted. We are a "Third Theatre" in Eugenio Barba's provocative term: the First Theatre being the national and state repertory theatres, the Second Theatre being the established avant-garde (Robert Wilson, Pina Bausch, Tadashi Suzuki and Robert Lepage). The Third Theatre is disenfranchised and emergent, those groups and individuals working on the edge, pursuing theatre more as a life-wish, as a vocation rather than a profession; those creating new work and forging new tendencies that are perhaps only acknowledged and absorbed into the mainstream many years later and often through a process of translation, mutation, and betrayal.

This is my work in the public domain as a producer, director, and curator. Within the academy – the university sphere – I am a teacher, an editor, a researcher. I specialize in contemporary theatre and performance, dramaturgy, and the theories and practice of non-Western dance/theatre. I am particularly interested in creative misunderstandings, the way in which Western theatre practitioners have "misunderstood" the concepts of non-Western dance/drama but elaborated efficacious theory upon such misconception. I hope that performance research and performance studies might remain unbranded, literally remain maverick but the threat of containment and institutionalization is coming nearer.

The Department of Theatre, Film and Television Studies at Aberystwyth was one of the first universities in the UK to develop an undergraduate program in performance studies, a formation nurtured and propelled by a particular ecology – the presence of CPR, the recruitment to faculty of Welsh and

Wales-based practitioners who were, and continue to be, established professionals in experimental theatre and dance, and the leadership of theorist/practitioner Mike Pearson. In an early course description, Mike wrote:

> Performance Studies provides an innovative, integrating and interdisciplinary perspective on the continuum of human action, from theatre and dance to public ceremonies, virtual performance and the performance of everyday life. But as performance has grown in importance as a means of exploring the myriad ways in which meaning is created and social life is shaped, it has been essential to introduce and include different histories, genealogies, geographies and politics into the fabric of the academic discipline and new practices and processes into the academic study of theatre.[2]

Now many universities worldwide have developed their own configurations of performance studies and ours is born from a specific set of circumstances forged and found in a very remote, isolated, and rural part of Western Europe and enriched through imbrications of theory and practice.

The value of confusion

Story Two. May 2000. I am entering New York, about to embark on a week-long external evaluation of the NYU Performance Studies program with Dwight Conquergood. The Immigration Officer asks me the purpose of my business. I state, "to evaluate NYU." He immediately responds, "Oh, so you'll need to be here several years then."

In a hectic week that sometimes felt like years, Dwight and I grappled with the extraordinary output of that department and its influence upon the field and I realize that what I value most about performance studies (from the NYU/Northwestern schools) is that it promotes a broad-spectrum approach to the appreciation and understanding of human endeavor – culture in the broadest sense. It is an optic, a way of looking at the world, and a way of constructing alternative views (proposing "new" realities, different options – making the familiar unfamiliar, opening perceptions).

Performance (as aesthetic practice) and performance studies (as academic discipline) together with performance research (as practice-informed theoretical enquiry) are burgeoning areas of development. Performance research is inclusive and intercultural, evolving and unsettled. It allows for new formulations and emphasizes process rather than product. It does not enshrine cultural values and pronounce upon them with certainty; rather it contests them and offers a space/site for dynamic negotiation. It includes uncertainty and diffidence, promotes experiment, nurtures a sense of becoming and evolving, and encourages reflection. It emphasizes the provisional, action with contingency, mutability – culture in a subjunctive mode.

It does not oppose, replace or deny the "Western" tradition of theatre (its canon and practice) rather it promotes this as one color amongst many in the spectrum and then positively embraces the performative traditions of other cultures, hybrid forms and innovative fusion. It encourages an understanding and appreciation of the methods, techniques, and aesthetic concepts of other cultures and societies both from around the world and within our own nation even as it confuses that notion. Perhaps that reach can be extended, perhaps there are very positive aspects to confusion in the way that there are also with regard to insecurity?

My own secure grasp on the world was certainly shaken by the Mappa Mundi in Hereford Cathedral. As a schoolboy at Hereford, I had to walk past it every day for seven years. This is the most extraordinary complete world map, made in 1242, that follows the medieval and classical construction of maps with a division of the world into Asia, Africa, and Europe, with Jerusalem in the centre. But by a terrible mistake at the end of the entire process, the cartographer misnamed the central sections – the east section (Africa) appears as Europe and the west section (Europe) – as Africa; it contains extraordinary depictions of places and ideas through the elaboration of a bestiary and a set of ideograms so when you find Crete, for example, you also see the labyrinth.

Last year to mark the thirtieth anniversary of CPR I used 30 objects to create a performance lecture. It was an exploration of memories, meanings, identity, and obligation, mediated through the enduring ability of objects to embody, invoke, and evoke the transitory and ephemeral nature of performance and history. Object 22 was a broken globe: the much traveled globe, made in the 1920s, its expanses of imperial pink receding into faded patchy pointillism, which finally fell apart in my hands, in rehearsal in Londrina, Brazil, latitude 20.23° south of the equator, the glue of the hemispheric seam, melting, yielding, cracking open, and fracturing at its point of weakness.[3] Evocative:

> People-to-people, community-to-community, artist-to-artist. A new internationalism, or an old one? The politics of exchange, the multi-edged ambiguities of "collaboration."
>
> Translation, mutation, betrayal, appropriation, exoticization, alterity, globalization, immunization, fundamentalism. East/West, North/South. The broken globe, breaking into more pieces.
>
> Meeting ... daring ... trying ... needing ... to touch ... there, right there ... the handshake, a moment ... there ... in the middle of the shifting fundament, the matrixed minefield ... there, in the no-man's land ... there, on the border, transit papers in hand ... there, in the sucking, smothering sludge of bureaucracy ... there ... for some ... on the other side of barbed wire and trenches ... We must.

> The broken globe, shattered geographies, the world in pieces, hemispheric shards dispersed into the cosmology. Humpty Dumpty who had a great fall – a fall into time and out of space – and all the King's horses and all the King's men who couldn't put Humpty together again.

Other values

I am standing in the circle transfixed by what I am seeing, shoulder to shoulder with other men all in white *djellabas*. The chants falling and rising, I am unable to understand the sequence of events unfolding before me. The ones inside the circle who are spinning, a young man who seems to be out of control, the dust rising, the old man in a green cloak, necklace upon necklace of beads, standing on the step on the cemetery surrounded by others, in a multicolored coat of rags, visited by many. This is extraordinary. I feel the same rush of excitement I felt many years before spending the entire night in the *Djemaa El Fna*, the Square of the Dead in Marrakesh, witnessing the finest range of storytelling, acrobatics, music, and magic I have ever seen. I am present at something intensely alive, realized in a frenzy, creating an ecstatic performance that makes all our talk of the performative seem insipid and sterile. I feel a hand firmly grasp my shoulder, I turn to the erudite and exceedingly polite professor of Arabic drama. We had met earlier that morning. He beckons me away from the circle. I am torn and leave reluctantly, anxious not to miss the next stages of the *Zikir* ceremony. Away from the circle now, the chants fading, he turns and asks in a rather conspiratorial tone: "Mr Gough, would you tell me please what is the latest play of Sheila Delaney?"

The situation is so awkward. I want to experience the *Zikir*. This is my only chance; I cannot be impolite. I know so little about English playwrights and the juxtaposition of situation to question borders on the surreal. But he does not relent and asks also of Ann Jellicoe. But to explain Ann's shift from Royal Court Theatre, and the development of the community play movement, in the shadow of Umm Durban cemetery with 500 men in the background participating in the most extraordinary ritual I have ever encountered, was utterly bizarre. The Professor had spent four years at Cambridge in the sixties and was anxious for news and in any case did not entirely approve of this Sufi ceremony. I had rejected so much of my early interest in plays and playwrights, proudly proclaiming in the late seventies to be "illiterate on purpose." Now I desired to be engulfed in a ritual ceremony that stretched my experience of performance.

Later, this led to an animated discussion about theatre studies and whether performance studies would be a better vehicle to describe and analyze the ritual discussed. Was this me at my proselytizing worst? As the then President of Performance Studies International, was I just another sort of missionary? Yet I would argue that performance is perhaps a better optic with which to view the natural/native performance of Sudan. Performance embraces diversity and difference, it places all that is on the periphery of Western Drama and Theatre Studies in the centre or indeed attempts to avoid the imperialistic notion of centre. It operates on the boundaries and borders of disciplines and militates towards an interdiscipline, eager to comprehend the performances of other cultures on their own terms.

But how does one know another culture?

Is to dance the dance, the *form*, of another tradition an indication that knowing, feeling, embracing another culture and understanding is any closer, nearer or more possible – in this day and age?

Theatre history describes how time and time again one highly developed, specific, peculiar, and idiosyncratic set of theatre practices and aesthetics has had formative and transformative influence on another nation's theatre culture – import, impact, importance. Is this a form of viral transmission? Does the export and import of theatre lead to contamination – infection in the most positive sense; a challenge to the old and the known, the tried and tested, a challenge to complaisance and tradition, inspiration for the new, an alternative view; an opening of the doors of imagination; an incorporation of new techniques and aesthetic practices – renewal, renaissance, and regeneration? And if so, where does that leave the culturally endemic, the native, the traditional, the specific, the local, and the located? What can the relationship be between innovation and tradition, the home-grown and the imported and what is lost when it is especially and purposely made for export: sanitized, packaged, and marketed. I think here of the export brew, the beer that is specifically made to travel and remain stable and similar wherever it is drunk in the world and yet the beer (or wine or theatre) we most want to drink and savor is the local, and a whole sequence of locals from around the world, that is specific and peculiar, of the terrain and *le terroir* (the sense and taste of place) – rooted, culturally specific; how do we satisfy these seemingly contradictory desires?

Defining objects

Story Three. I am leaving Providence, Rhode Island, with a colleague in April 2005, trying to check four huge bags through to Manchester and carry two more on as hand luggage containing a bizarre collection of props, costumes, and culinary instruments taken as "contingency" for the devising of an opening event for the eleventh Performance Studies International (PSi) conference at Brown University. The contents alarm the x-ray operator and a thorough manual inspection reveals large quantities of syringes, test tubes, test tube holders, pipettes, and aluminum flasks. "Are you a scientist Sir?" Thinking that conforming to the occupation and the research that the objects suggest and represent was the easiest solution to an otherwise complicated explanation, I responded affirmatively – "YES," I say, "I am a scientist." It sounded good, I was convinced.

Moving on to the next line of homeland security, the contents of one bag cause the scanner even greater alarm. Rapidly discovering the inoffensive but stainless steel implement the security guard begins to restuff the bag. He looks up to me and says, "Sir, I couldn't help but notice the wigs – what's with the wigs?" Desperately speculating whether this was an offence – could you be in big trouble for cross-dressing, mid flight, or disappearing into the plane's toilet for a make over? I stumble for words. He says, "I notice they look theatrical and old-fashioned." I am thinking, do they have some sort of sophisticated cross-reference system going on here? If I told the guy at check-in that I was a scientist, can I be a Judge at the security gate? Again, I place confidence in the significance of the objects and the network of meanings they promote. I look at him sternly and say: "YES, I am a judge." I have often found that the best way to cut short talkative and prying taxi-drivers is to say that I am a dentist, it ends *all* conversation. My best roles are played out in everyday life; occupations are mutable, and easily inhabited.

Local values

He is putting his hand into the frying pans; all six burners are alight, he is moving ingredients from one pan to another, tenderly, sensually, efficiently. We witness craft, skill, and virtuosity but it is not showmanship; we see it as performance, as choreography, as art, but this a process of transmutation by knife and fire, cutting and cooking not for display, not for visual effect but primarily for the impact on taste and smell. Of the many extraordinary culinary events witnessed and consumed during the conference, Points of Contact: Performance Food and Cookery (1994), the demonstration by Franco Taruschio, Wales's foremost chef, was the most beguiling. I had invited him to give the antidote to all celebrity chef cookery demonstrations: to work in silence without explaining or "talking up" the process and to make, with two assistants in a demonstration kitchen, his favorite three-course meal. One hundred people watched mesmerized, dribbling at the mouth. This moment, and the entire conference, has influenced my work over the last ten years in making performance work that functions on the border between edible installation and performative banquet, table and stage, participant and witness, forager and fed.

The conference explored food in performance and food as performing art; the performative in cookery, its staging in the kitchen and at the table; the processes in cooking and performance making, exploring piquant analogies and correlations; the theatricality of food and food as a model for theatre, multisensory, processual, and communal. We wanted to bring together artists, chefs, and scholars from several cultures in an interdisciplinary collaboration. And we promised that participants would enjoy an immersion experience at the intersection of food and performance, theory and practice.

One of the many strands of the CPR program, Points of Contact, is a continuing series of international conferences that aims to generate investigation, critical debate and cross-disciplinary approaches to performance research through contact and confrontation with other disciplines. The first seven in the series have included: Performance, Politics, and Ideology (1990), Performance, Ritual, and Shamanism (1993), Performance, Tourism, and Identity (1996). Future themes include Performance, Homes, and Gardens; and Performance, Health, and Medicine. These conferences have contributed significantly to the development of a distinctive approach to performance studies in Wales coupled with a rigorous integration of theory and practice, an emphasis on aesthetic practice, an interrogation of practice through theory (and vice versa) and a sustained effort to collaborate with other disciplines, specifically Geography, History, Earth Studies, Landscape Studies, Archaeology, and most recently with our prestigious Department of International Politics at Aberystwyth University. Much of what has been achieved in this interdisciplinary approach is brokered through a set of located practices and specifically Welsh perspectives.

In terms of contemporary theatre practice, the innovative is championed and promoted in Wales by the CPR and a consortium of practitioner/scholars; we pioneer and create the opportunities for the new, the unpredictable, and the unforeseen. But we have always said that we are interested in the relationship between innovation and tradition and we encourage translation, mutation, and betrayal with regard to methods of, and approaches to, performance making.

The CPR's role in Wales is as a "conduit" for exchange, co-production, and contrast, to investigate and nurture emerging artists and ideas, and at the same time to pursue – sometimes against prevailing trends and fashions – the questions underneath, in ongoing long-term research; inscribed within every project, every activity mounted by CPR is the desire to effect change, with key objectives being investigation, discovery, and sharing: to bring these influences to bear in Wales, where a "new theatre" is possible, and can be formed distinct and whole without reference or derivation from the English mainstream model.

This motivation connects and contributes to a wider cultural, social, and political agenda. Wales needs to determine a place in Europe, distinct from England, and distinctive.

Philosophically and in practice, however, the CPR has no desire to falsely celebrate national culture; it gravitates towards the reflective, preferring to raise curiosity, to question, and to challenge. Whilst supporting initiatives that make strong links with countries/nations of a similar dimension, the CPR also believes that people and nations develop by engaging with difference; recognizing otherness and determining one's own unique characteristics through dynamic interaction.

Wales as a work in progress
As a project of imagining and re-invention
As a process of collaboration and co-operation
As a place of inclusion and integration
Where a sense of nationhood is determined through dynamic negotiation and transaction
Where diversity is celebrated and cultural pluralism enjoyed
Where the "script" is not authored by an individual but devised through collective action
Where the ending or "endings" can be altered or changed
Where the participants determine the outcome and a sense of ownership is gained
Where new notions of art and cultural action are integrated with new technologies and a new nation is proposed

Notes

1. Note: This is an edited transcript of a spoken text.
2. From an unpublished course description by Mike Pearson (1997): http://www.aber.ac.uk/~psswww/pf/scheme/introduction.htm.
3. Editors' note: The "expanses of imperial pink" refers to the colonial practice of using pink to depict England's colonies on the world map.

4
Using Performance as a Practice as Research Tool in Africa

Jane Plastow

This chapter will focus on two Arts for Development projects in the Horn of Africa; their contribution to my thinking on the ways performance can empower human development; and how I have sought to bring this practice into my teaching, most notably in the Masters course I run at Leeds University in the United Kingdom in Theatre and Development Studies.[1]

The arts have been used within development contexts in Africa since the 1970s,[2] and today there is undoubtedly more funding for development-based performance work across the continent than there is for conventional professional companies. However, the funding agencies – international non-governmental agencies (INGOs) such as Care International, Oxfam, Save the Children, etc.; UNESCO or government bodies such as ministries of education and culture – seldom have experts in Arts for Development and often commission work cheaply from local groups to support worthy initiatives – such as the promotion of understanding about HIV/AIDS, or underscoring the importance of education or supporting women's rights – without thinking about how any particular project can best work with a target community to empower it to take control of the issue in question. The common pattern is that a group of enthusiastic, but minimally or totally untrained school leavers with an interest in drama is commissioned to put on a play, often in a market space or at a community gathering, in order to provide the information the funding agency wants conveyed. Considerations of form, follow-up, facilitation, and sustainability, to name but a few, are seldom considered by the agency, which often judges success simply by the number of people who have witnessed the production.

This extraordinary situation – would you send someone to act as a dentist who had a keen interest in oral hygiene and could read an information pack? – is of course being challenged across the continent by many skilled professionals. A number of African universities run courses in Theatre for Development, and at Leeds we have been working with African academics and practitioners for some 40 years looking at the social uses of theatre.[3] There is also now a substantial literature on Arts for Development.[4] The problem is that, firstly, issues such as women's and children's rights, or using the arts not just for information but for real empowerment, which had previously not taken centre stage, have come to the fore and require new thinking about how best to take work forward. Secondly, development and arts professionals have seldom worked together as equal partners, with development still being dominated by an economics rather than a participatory human-centered approach, and development professionals seeing the arts only as a useful sideline in which they need no particular expertise.

My own work, then, has focused over more than 20 years on experimenting with three main aspects of Arts for Development. How do we best train local people to use the arts effectively and sustainably with their own communities? How can we get development agencies to understand and engage with the true potential of arts as a means of empowerment? What techniques and strategies can we best use to make performances that are beautiful, meaningful, and useful to the communities we work with? I am happy to embrace a label I was once offered by my co-devisor of the MA in Theatre and Development Studies, Professor Ruth Pearson, of being a "pracademic" – one who uses practice centrally to develop theoretical and academic thought in their subject area.

Ethiopia: The Adugna Dance Theatre Company

> They didn't only teach dance, they gave us life. How to live and change others. This is life and what Ethiopia needs.
>
> (Meskerem Tadesse, Adugna dancer[5])

Between 1997 and 2001 I was privileged to be asked by a local non-governmental organization, the Ethiopian Gemini Trust, to assist them in developing community outreach strategies, Theatre for Development training, and evaluation of their program with the Adugna Dance Theatre Company. Adugna then consisted of 18 dance trainees drawn from the street children of the capital city of Addis Ababa. The program was run by a British dance trainer, Royston Maldoom, and existed to provide professional training for some of the children Gemini works with, to raise the profile of impoverished children, and to train a group of young people to act as advocates for their own community using dance as their primary medium.[6]

Adugna was hugely ambitious. It took young people with no idea of dance training and aimed to make them into Ethiopian and contemporary dance specialists. Originally the intention was to work for 18 months, but this gradually extended to a five-year period, because the young people involved needed not only dance training, but also a whole range of new skills. Life on the streets is violent, aggressive, and often drug addicted. It is also extremely short term. In the first workshops I ran, I asked children to show me their dreams; these extended only as far as having a decent meal or some petty goods to trade. The project's backers wanted the group to understand the issues behind the problems experienced by street dwellers, to be able to act as dance trainers for others, and as advocates for their community. This involved not just technical training but human development, and that takes time, trust, and support.

The major funders for the project were a UK organization called Comic Relief. I wrote three evaluations for the agency, but maintaining funding was an uphill struggle. In essence, the agency was dubious about putting substantial amounts of money over a number of years into the training of only 18 young people.[7] For me this crystallized an issue I had been struggling with: does the development industry want to ameliorate poverty or to transform the impoverished into fully equipped agents for change? Too often it seems the aim is merely amelioration that will maintain a quiet, grateful group of subordinate people, and that this is in essence a neo-colonial agenda. Transformation is much more challenging and requires long-term sustainable input. It also requires listening properly to the client group.

As the young people became more confident they also became more demanding. They started asking for financial support to help their families, the women became much more assertive, and they challenged some of the decisions Gemini staff were making on their behalf. Some saw this as ingratitude: as going against the hierarchical, patriarchal values of local Amhara society, but I came to see it as a necessary and healthy process of beginning to value oneself. Docile gratitude is never going to lead to profound change, but truly participatory development, such as the Adugna program, asks people to start thinking and acting – and probably to start challenging authority.

The other main point of learning for me in this project was the use of dance itself. In African performance, dance is often integral to any production, and I had often used it in plays and training programs, but here dance was the main medium of expression, especially the introduced form of contemporary dance. I am very interested in appropriate use of form, and usually use people's own performance forms as much as possible so as to make work accessible and meaningful, but here the majority of dancers said they preferred contemporary dance. This seemed to be because the indigenous dance forms as taught by Ethiopian teachers required learning the correct way to perform existing dances, but little was discussed of their meaning or contemporary relevance, whereas contemporary dance could be molded to deal with any subject matter. The dancers felt they owned and could choreograph contemporary dance, even though it was an imported form, more readily than they could pre-existing Ethiopian dance forms. Artforms that are not malleable simply become museum pieces, and are often socially restricting, inscribing old patterns such as the inferiority or sexual availability of women.

The unresolved challenge here seemed to me how we might use some of the patterning and technique of local dance forms in new or radical ways.

What became clear through the dance was just how much one could learn through the body. Instead of cowering in corners as some had when we first met, the dancers came to learn the power of taking space – on the dance floor and in life. They also learnt trust and support, as dancers working together, lifting and leaning on each other. This embodied learning seems to have gone further with the group than any amount of words could have done. In their art and practice, Adugna work with each other and the community groups they serve; their self-confidence comes from the extraordinary high-level training that now allows Adugna to dance at international festivals, to train the Ethiopian police force in dealing with the street community, and to run free dance classes throughout the slums of Addis Ababa.

Eritrea: arts and education

The pilot program I ran in Eritrea in December 2005 and January 2006 built on an engagement with arts in Eritrea that goes back to the country's independence in 1991. However, it was a new direction for me, working with an education colleague, John Holmes, to use theatre to identify what children thought about the education they were receiving. The context for the project was that the Eritrean Ministry of Education was trying to move away from a teacher-led to a more child-centered teaching model and was also planning to incorporate arts work for the first time in the primary school curriculum. The plan was that while Holmes and colleagues from the teacher training college ran in-service workshops on child-centered learning, I would work with my long-time Eritrean collaborators, Mesner Andu and Yakim Tesfaye, with groups of final year primary school students to discover what they thought of the education they were receiving.

I will concentrate here on my experiences in one particular school, working in the local language of Tigrinya, in the mountain village of She'eb. We worked with the school for only one week, because this was intended as both a fact-finding exercise and an experiment in just how open and engaged the children would be with the arts process.

The 30 children we met were extremely shy. Foreigners come seldom to She'eb, the culture expects children to receive quietly rather than to question, and arts work in an education context was entirely new to the group. We began with a small explanation of what we were doing before moving on to games and then asking the children to draw images of what they liked and disliked about their school experience. This resulted in extreme concentration and a range of images of the Eritrean flag, which is flown outside every school and symbolic of strongly inculcated nationalist feelings; friends and particular classes – Tigrinya and math seemed most popular. When children were asked to draw what they didn't like, they were initially hesitant since young people are not normally expected to criticize authority, but they came up with images of dirt around the school, of children fighting and of bullies, and of classes they didn't like – English was top.

We all discussed the pictures before moving on to making sculptural images. Once more the focus was on difficult classes and friction among children, but the facilitation team noticed there was violence in many of the images: teachers hitting pupils and children fighting. When we discussed the images with the children they did not focus on adult violence, but rather on the problems they had with learning. When we moved on and asked the groups to show us their ideal images of resolution, we saw plenty of images of children helping each other – but only in one case was a teacher shown in a helpful role.

That evening, we discussed our findings with the group led by John Holmes. It emerged that issues of punishment are seldom discussed in teacher training college. As in most of Africa, corporal punishment is accepted, but the idea that children might value encouragement seemed scarcely to have been considered. Both groups decided that they wished to pursue this question further, and the next day we asked the children to explore making small transitional scenes, showing how they might move from the problem to the ideal. Again, it appeared that only children could be looked to for help. When I intervened to show a teacher being supportive, I was met by slightly incredulous giggles.

Now that the children were becoming more relaxed, my Eritrean colleagues led group discussions about issues of child-on-child violence and adult-on-child punishment. What emerged was a scenario fairly shocking to all the adults – though seen as normal by the children – of a range of excessive physical punishments, including beatings, being made to walk on one's knees across a stony playground, and a bizarre punishment that involved pens being put between the fingers and then the hand being twisted around them. The playlet we developed to show teachers, fellow pupils, and interested parents was based on stories the children told us about these punishments for arriving late at school and teachers' failures to enquire as to whether there was a good reason for lateness.

What was of interest to me as a process was, firstly, that we had come up with a topic none of us had anticipated. In schemes where an agency says what a play is to be about, this potential for learning what the real issues of concern might be is pre-empted by the funders' or playmakers' beliefs that they know what is important. Secondly, the issue had only emerged through a multi-arts approach in which we used drawings, songs, sculptural images, and play making. Various children engaged more with one means of communication than another, but discussion only became possible after issues had been raised in a non-verbal form. For many oppressed people it is impossible to simply come out and say what problems exist and, as here, some oppressions have become so much part of the fabric of life that they are not seen as matters to necessarily complain about. Only a gentle and circuitous approach allowed an issue that challenged existing authority to be raised. Finally, the response of the adults when they finally saw the play was not condemnatory. Some teachers did say there had been exaggeration, but mostly they were interested in discussing the issue and were fascinated to think that pupils might value their encouragement. I should also mention that education officials who saw the production subsequently investigated the school. It was found that the head teacher was corrupt and had been stealing resources; he was replaced. In the other school we worked in there was nothing like the violence we encountered at She'eb.

Using practice within UK teaching

I teach Theatre for Development and Theatre of the Oppressed as part of a core course on the uses of theatre in the MA I run. This usually has a very international student body, most of whom already have varying degrees of practical experience in making community-based performance. The teaching combines theoretical and case study reading with discussion and viewing of video recordings plus workshop sessions. I could not imagine leading this teaching if I had only a theoretical understanding of the issues involved. I commonly use my own recordings and some of my writing within the course and discussions draw on both academic reading and the lived experience of those taking part in the course – including my own. I draw on my practice as research in discussions with a number of PhD students I have working on arts for development topics.

For me, practice, teaching, and research are part of a single whole, which I find quite difficult to separate out. This is often problematic within the academy. I work within a large School of English and while interdisciplinarity and practice as research (PaR) have now come to be accepted as valid terms, some colleagues still see the most prestigious research as involving the solitary writing of monographs. The fact that much of my writing, like my practice, is collaborative, has led to further difficulties historically in gaining recognition for the value of my work. Moreover, the nature of my experimentation within the context of development issues puts me in a different position from many PaR practitioners who are exploring less applied arts approaches. However, my theatre colleagues certainly value the work, and I commonly travel to talk and run workshops across a variety of arts and social sciences disciplines in the United Kingdom and abroad. Answers remain elusive, formulae non-existent, but the questions about how to make the next piece of work with the next particular group are always fascinating, and trying to educate new practitioners coming through the Masters course, who will understand the discourses of both theatre and development, seems to me the best way to provide a receptive soil for best and widespread practice of Arts for Development in the future. Yet if I undertook this training without continuing my own practical explorations, and without continuing to reflect on what I learn from each project, I am sure my teaching would soon become sterile and out of touch. PaR is probably what defines me in any way as a useful academic and allows me to follow the elusive dream of using art and learning together to transform the world into a creative, thoughtful, and infinitely enjoyable space where ordinary people have agency over lives which they and others can recognize as valuable.

Notes

1. The MA in Theatre and Development Studies is jointly taught by the two departments concerned and is one of only two such courses internationally – the other at the University of East Anglia, UK.
2. It is commonly accepted that Theatre for Development (TfD) has links back to Laedza Batanani, a Botswanan initiative with agricultural extension workers in the 1970s.
3. The Workshop Theatre at the University of Leeds set up the first MA course in Theatre Studies in the UK (1967). It has for many years run a popular and influential course on the social uses of theatre and has had strong links with Africa since its inception.
4. *See* Richard Boon and Jane Plastow, eds, *Theatre Matters: Performance and Culture on the World Stage* (Cambridge: Cambridge University Press, 1998); Eugene van Erven, *Community Theatre: Global Perspectives* (London and New York: Routledge, 2000); and Jumai Ewu, ed. *Theatre for Development: A Digest of Experiences*, vol. 12 (Contemporary Theatre Review, 2002).
5. From an interview with the author in 2001.
6. *See* Jane Plastow, "Dance and Transformation: The Adugna Community Dance Theatre, Ethiopia," in *Theatre and Empowerment*, ed. Richard Boon and Jane Plastow (Cambridge: Cambridge University Press, 2004).
7. The cost of the training project was estimated to be £225,000.

5

Rating the Theatre Practitioner: A South African Case Study[1]

Temple Hauptfleisch

The debate on practice as research (PaR) in South Africa shares the same sense of frustration experienced internationally. This feeling has grown stronger over the years, particularly with the advent of a new, highly pragmatic, *outcomes-based* approach to tertiary education and a controversial system of financial reward for research outputs.

Numerous meetings and conferences have addressed this topic, but with limited impact on the larger system(s). Most influential in theatre have been a series of recent biennial conferences named Dramatic Learning Spaces, which led to an ongoing PaR working group, a national Peer Review Project (PRP), and the 2007 IFTR Annual Conference in Stellenbosch.

Though *formal, national* acceptance still eludes the country's artist-teachers, these initiatives have had some effect, since almost all Universities now recognize PaR to some extent and have (internal) recognition system(s) in place for creative outputs and processes as research.

The local debate was fuelled by the unique rewards system for *research outputs,* introduced in the 1980s by the Department of Education (DoE) as part of its tertiary funding formula. Aimed at encouraging academic research and publication, this scheme pays institutions a *specific amount per output unit* produced by their academic staff. (Output units are classified in five categories and are only awarded for publication in journals accredited by the DoE.) Since many institutions pass (part of) the money on to the particular department or individual researcher, this becomes a source of considerable additional research funds for prolific writers.

Unfortunately, the program has never recognized creative outputs as the equivalent of formal articles or books, and adamantly refuses to do so. Two strong and compelling arguments are made for their exclusion: (1) the process of making art is an autonomous activity with its own unique infrastructures and funding and reward systems, and (2) it is difficult to obtain peer reviews of outputs.

The National Research Foundation (NRF) rating process

In the 1980s, the South African NRF introduced a rating system for natural scientists and in 2002 extended it to the social sciences and humanities. Intended to advance the quality and quantity of research at tertiary institutions, it uses a national and international benchmarking system that focuses on individual researchers rather than institutions (as is common elsewhere). Utilizing peer review and various review and appeal processes, it places applicants in six broad categories, based primarily on their active output of the previous seven years:

A: A leading *international* scholar
B: Someone with *considerable international recognition*
C: A leading *national* scholar
P: Young researchers (normally younger than 35 years of age) with *exceptional potential.*
Y: Young researchers with the *potential to establish themselves* as researchers within a five-year period after evaluation
NR: Not rated, i.e., *does not qualify* for any of the categories.[2]

Generally welcomed, the extension was nevertheless controversial because of a perceived lack of consistent and "objective" evaluation and assessment criteria for the humanities. To their credit, the NRF included the arts (creative arts, performing arts, and design) in the process and altered their definition of research to include conventional research (for example, the history and analysis of theatre) as well as artistic output of staff and students at tertiary institutions.

Stating that *research* refers to "original investigation undertaken to gain knowledge and/or enhance understanding," the definition adds two new items, namely: "the creation and development of the intellectual infrastructure of subjects and disciplines (e.g. through dictionaries, scholarly editions, catalogues and contributions to major research databases)" and "the invention or generation of ideas, images, performances and artifacts where these manifestly embody new or substantially developed insights."[3] The significance is that a strategically placed governmental institution now recognizes artist-researchers, journal and book editors, curators, encyclopaedists, and archivists, *as researchers*.

Key issues raised by the rating process

Predictably, the *application* of these principles to the arts has not been easy. Let us consider some core issues foregrounded in the initial stages of implementation.

The notion of "research" in the arts

It seems to me that there are four possible approaches to the notion of research in the arts:

(a) Arts research as *the study OF the arts.*
The object of study is the work of art, the methods are rooted in arts theory, and the research output is a read paper or a written report, presented at conferences, in journals, books, and electronic format. This is, the conventional approach used by literary, dramatological, musicological, and art-historical studies and part of the venerable tradition of written scholarship. There are no reward and funding problems here.

(b) Arts research as *a study undertaken THROUGH/BY MEANS OF the arts.*
Here the object of study is some issue (for example, in society), the methodology is the process of making the work of art, and the output is, for example, a performance (and/or a published text). This constitutes a unique form of "soft science," where the work of art itself is seen as simultaneously "process" and "research output."
In the words of John Gardner, "When fiction becomes thought [. . .] the writer makes discoveries which, in the act of discovering them in his fiction, he communicates to his readers."[4] This notion lies at the heart of the international PaR debate. Unfortunately, few conventional research institutions see this as legitimate, unless some *other* person (a critic or scholar) *interprets* the work *in a written and formally published document.* While the creator/performer could take on the multiple roles of creator, observer, *and* researcher, few artists actually want to write conventional reports (or feel competent to do so), and thus tend to insist that the performance must be both the *thing studied* and the *report about the study.*

(c) Arts research as *the DEVELOPMENT OF NEW TECHNIQUES AND PROCESSES for making art.*
Making an artwork is interpreted as a form of *developmental research,* in which the range and nature of an artform is expanded. An accepted core idea in the natural and industrial sciences, it is considered problematic in the arts, *unless a more conventional report is part of the process.*

(d) Arts research as *the development of NEW TECHNOLOGIES AND INSTRUMENTS for use by artists.*
Related to (c), but not problematic, is the development of new technologies (for example, new paints or lighting systems) for producing art. This process and the registration of patents is a standard and accepted process of reporting in the natural and industrial sciences.

Defining and rating "research outputs"

Complex as the process of arts research is, the issue of what constitutes an acceptable *outcome* or *output* in the arts is even more problematic. For millennia, the prime means of reportage has been the written word and the notion that the artwork could be both process and report has been consistently rejected. In this respect the NRF's new definition, outlining a broader, generic notion of what may constitute "outputs" in the arts, is of particular importance. Using this, the NRF Panel developed a working document containing a list of acceptable types of research output (with examples). Included were scripts for, performances in, and the direction and design of live and recorded presentations, "provided they can be shown to have entered the public domain and manifestly embody new or substantially developed insights."[5]

The statement was a crucial and useful criterion for distinguishing between performances, which entail a *specific research aim* and distinguishable *research outcomes*, and the bulk of performances, which do not. Substantial problems arose, however, when the panel had to assess the *research* component involved in applications from teachers, performers or technicians, individuals whose contributions tend to be subservient to the performance as a whole. Consider two examples:

1. A *playwright* writes a play to explore a social issue and the play is publicly performed and/or published. Here playmaking is a *research process* and the findings are provided *in the play*. A similar argument may be made for a *director's* exploration of themes in an existing play. These examples clearly fit the new, expanded definition of research.
2. A *performer* plays the leading role in one of the above. Is this "research"? The usual answer is no, for s/he is basically *interpreting* and filling out a character as set by the author and/or director. The argument may be that research is required to "find" the character – but what precisely is the research *outcome* here?

Improvisational and group work in theatre is another matter of course, for the *ensemble* now becomes the creator, with the performance a group product. Thus everyone involved can lay claim to it as team research and the performance as *their* (joint) output – as long as supporting materials document the process.

This raises the question of *replicability* and the need for some kind of supportive material.

The replicability of research

The notion of *replicability* is universally accepted in the evaluation of research output. In theory this should not be an issue for serious theatre practitioners, since they look at other people's work not to repeat it, but to do something that has not yet been done.

Unfortunately, most assessment procedures do not focus on *process,* but on the final *product*, and "replicability" suggests that any output must contain, *within itself*, a description of the aims, the methods, the process(es) utilized, as well as the result(s). The peer reviewer requires this information to evaluate the output/results against the aims, intentions, and methodology set by the researcher/artist, for without it only the *result* can be judged, the rest merely intuited.

The conventional proposal is that the researcher/artist provide the peer review panel with a *written* outline of the process. (Perhaps a program note, catalogue, or video or taped discussion.[6]) This minimum requirement is set for all conventional researchers, but it is often vehemently rejected (also at meetings of the NRF Panel itself!) as a discriminatory imposition that denies the fundamental nature of the communication processes in the arts. The argument is that *the work must speak for itself*, communicating its aims, its processes, and its results in one presentation, or performance. This argument remains unacceptable to conventional researchers and research administrators, who see it as a strategy for privileging the "artist-as-researcher" above any other.

Millie Taylor suggests that artist/researchers develop "[a] new form of presentation/performance/documentation [...] to enable understanding of other stages of the investigation and the conclusions"[7] – one perhaps more acceptable to the research establishment. Logical as it sounds, the idea is unlikely to appeal to most artists, for at heart they want to be *seen* and *treated* as researchers, but not to *think or act* like them. They want to share in the privileges of researchers (grants, sabbaticals, rewards), but on their own terms as "artists," retaining traditional privileges *denied* scientists (arts grants, arts awards, and income from artworks).

This is perhaps the most crucial hurdle to overcome if the serious artist/researcher is to be recognized, not only by the rating system, but also by all other awards systems. The key seems to me to lie in the nature of the contextual discussion (and motivational supporting materials) required, rather than in the fact that it *is* being required.

Peer review

Despite its own perceived fallibility, peer review is the most broadly accepted principle for the assessment of academic excellence and the credibility of the academic enterprise. It is thus also one of the most troubling questions in the call to recognize creative outputs. How is it to be done? *Can* it be done? Who is to pay for the undoubted expenses to be incurred in evaluating ephemeral performances and site-bound exhibitions spread across the globe? How is the very range and variety of the arts to be comparatively judged?

Peer review of conventional research works because the existing systems of journal and monograph publication *depend on peer review themselves*. By contrast, much of the work in the arts is "published" in non-conventional ways and most of it is *not* subject to conventional forms of peer review, though some may be.

None of the existing bodies (tertiary institutions, NRF, DoE), view peer review of the individual output as a flexible matter, and certainly not as something which can simply be waived in the case of the arts. Indeed, they consider the lack of a system of peer review (and the difficulties of actually creating one) another *key* stumbling block in considering arts outputs as research outputs.

To this end the NRF took a proactive step in 2005 by funding an experimental project on peer review, in which a national project team, led by Mark Fleishman of the University of Cape Town, sought to develop a national peer-review system for theatre, with a first round of trial evaluations in 2006, followed by an assessment meeting and a second round in 2007. The results are currently being evaluated and written up.[8]

Referees

Besides utilizing *existing* peer review in the various disciplines, the NRF rating process also has its own peer-reviewing processes and presents its own particular problems.

Every application goes to a number of *peer reviewers*, whose reports form the basis for rating. The NRF Panel, which consists of one specialist per broad arts discipline only, considers and interprets the referee reports, not the applications themselves, while the administrative staff and management of the NRF do not comment on applications.[9] Obtaining six competent referees per candidate is enormously important, but extremely difficult given the small pool of academic artists in the country. This is complicated by the difficulty all panels have in obtaining good and useful responses – or even any response at all.

Five factors seem to play a crucial role:

(a) **Unfamiliarity with the rating process and the role played by peer review of the art work(s).** Creative artists often find it difficult to judge another colleague's work *as research* and write an assessment, which is understandable to non-specialists. Even experienced critics and historians often provide an evaluation of the person and his/her impact, rather than of the *quality of the particular work*(s).

(b) **Personal bias.** This can be overly positive or negative. The former may lead to blindly supportive comments on *any* candidate, irrespective of the merit of the work. Negative bias is common in both the small and specialized worlds of the arts and academia, ridden with internecine strife, ideological differences, petty jealousies and so on.

(c) **Rejection of the rating system per se.** The whole system is in its infancy (as far as the social sciences and humanities are concerned) and not everyone has welcomed it. When approached as referees some people attack the credibility of the system or refuse to participate.
(d) **Overloading of available academics.** This is especially true of those able to write a referee report of substance.
(e) **Unfamiliarity with the person and/or work.** International referees in particular have a problem here. Not having seen the live work, how are they to judge? The key problem is that so much of the work in the arts is ephemeral or too bulky (or tied to a context) to be moved.

What is required is a working peer-review system producing critical, well-substantiated and clearly communicated referee reports, by referees who accept, understand, and support the process – and receive rewards for participation. Only this will make the system work and find general acceptance.

Quantity versus quality and impact

Not always well understood is the *difference* between the DoE rewards system and the NRF rating system. The former is simply a (mechanical) way of rewarding a researcher for having produced a piece of peer-reviewed work – irrespective of merit or impact of the output. Valuable as the process has been to the expansion of research at tertiary level, it is a rather blunt instrument, not primarily concerned with raising the quality of research, but merely with increasing the volume of output.

The rating system has the opposite aim; primarily focused on the stature and impact of the individual *as researcher within his/her field* and thus the *quality* of the work – the originality of the process(es) and findings, the impact and influence on future research/art production – is of paramount importance.

Most other processes of recognition, appointment, and promotion at academic institutions tend to be influenced by the same two opposing (though not un-reconcilable) imperatives: to appoint, promote, and/or support the proficient (income-producing) "hack" or the highly rated (and prestige-producing) "genius."

Facilitation of research

Really requiring a substantial article in itself is the question of "research facilitators" – those editors, publishers, facilitators, archivists, producers, curators, and compilers who drive the peer reviewing processes and publish, display, or present the outcomes. Totally denied in many evaluation, reward, and rating processes both nationally and internationally over the years, their work, ironically, informs the whole notion of a *research community* and the very *academic system* itself – for both presuppose the existence of a specific infrastructure, set up and maintained by people whose sole role may be to *facilitate* the production of research outcomes by others.

Neither the DoE nor the NRF yet acknowledge this in their systems of recognition (though – to be fair – most academic institutions do value more conventional activities such as editorialships and book compilations, if only internally).

International exposure

Finally, the emphasis on international stature means that no one whose work is not known and appreciated *outside* of the country is rated higher than a *C*. In a system utilizing an international benchmarking system this makes sense, though each subject in the social sciences and humanities can make a case for the regional/local nature of some aspects of its studies (for example, site-specific performances of San narratives), which do not "transport" easily to international interest and understanding. More critically, anything *not written in English* is automatically disqualified from categories A and B, since few international peer referees understand any of the other 11 national languages in South Africa, some which have produced a substantial body of academic literature.

The NRF remains adamant that the international dimension makes A-rated researchers so sought after; therefore the onus remains on artists to devise creative ways to place their work in the international arena.

Conclusion

The NRF's rating system is only *one* of many recognition and reward systems in place for the artist in the academic community, but it is a particularly interesting and positive initiative by a scientific institution keen on expanding its own horizons and enhancing the *quality* of research in South Africa. This in itself has gone a long way towards revitalizing the PaR debate in Southern Africa.

Notes

1. Founded on the author's experiences as a founding member of the NRF Rating Panel, some of these arguments were first presented at *Dramatic Learning Spaces: A South African Research Conference* in Pietermaritzburg on 24 April 2004 and published in *South African Theatre Journal (SATJ)* 19 (2005): 9–34. This article explores those early ideas further.
2. NRF, "Definition of Rating Categories," National Research Foundation, *see* http://evaluation.nrf.ac.za/Content/Documents/Rating/ratingcategories.doc.
3. Ibid., "Definition of Research," National Research Foundation, *see* http://evaluation.nrf.ac.za/Content/Documents/ Rating/definition_research_2005.PDF.
4. John Gardner, *On Moral Fiction* (New York: Basic Books, 1978), 107–9.
5. Temple Hauptfleisch, "Artistic Outputs, Arts Research and the Rating of the Theatre Practitioner as Researcher – Some Responses to the NRF Rating System after the First Three Years," *South African Theatre Journal (SATJ)* 19 (2005): 22.

6. Interestingly, the NRF definition seems to favor this idea and opens up possibilities for the inclusion of substantial illustrated program notes as part of a production, when they accept an *art exhibition catalogue* as a research output.
7. Millie Taylor, "Schrödinger's Research: When Does Practice as Research Occur?," *South African Theatre Journal (SATJ)* 19 (2005): 188.
8. Based on NRF, "Report on a Workshop on Research in Drama, Theatre and Performance in South African HE Institutions," UCT Drama Website, *see* www.drama.uct.ac.za.
9. On the rating process, *see* the NRF website: http://www.nrf.ac.za.

6
Performance as Research in Australia: Legitimating Epistemologies

Brad Haseman

It is a curious and perhaps surprising observation but the performance as research movement in Australia was most active, and arguably did its most groundbreaking work, nearly 20 years ago. In the late 1980s a shakeup of the higher education sector in Australia saw the amalgamation of smaller Colleges of Advanced Education into a national system of 39 universities and, along with this, grew heightened expectations around research. For many staff of these Colleges of Advanced Education, this was a time of both opportunity and nervousness. Few felt prepared for the research challenges which lay ahead. In the new order academics were to successfully compete for external research funds and build a strong staff and postgraduate research culture. This marked a major shift of priorities, especially for academics in the creative arts, media, and design disciplines whose principal focus had been on professional and vocational training and undergraduate course innovation.

By 1991 ADSA[1] was one professional body prepared to lead the way into research for academics in drama, theatre, and performance. From 1991 to 1995 they undertook a rigorous and reflective exercise, testing the performance as performance as research approach and developing a serviceable framework to establish it as a central feature of university research. The first part of this chapter deals with their proposals to define and shape performance as research at the time and demonstrates how ADSA was poised to influence the national debate about creative practice as research. The concluding section notes how ADSA's advances became missed opportunities for the field and outlines how performance as research is being framed for the contemporary environment.

Groundbreaking first principles 1991–1995

While colleges and institutes were being amalgamated, ADSA formed a Performance as Research Subcommittee,[2] which was quick to see the importance of embedding in the national research landscape, a research methodology congenial to their needs as scholars and practitioners. This was a direct response to their working environment, where performance practice in colleges was understood as a means of imparting skills and building the professional practices students would need upon graduation. This was particularly so in actor training programs and courses which prepared drama and dance teachers to meet the increasing demand for the performing arts in primary and high schools. Performance making was frequently seen as "an 'added extra'; something to make the text live, something to give students what was vaguely called 'experience.'"[3] Such attitudes reinforced a "strict division between practice (the proper province of practitioners) and description and analysis (the proper province of the scholar)."[4]

Many sought a way around this situation in which the playmakers and theatre practitioners who actually created the objects of study were asked to stand aside and leave the task of analysis and interpretation to the "proper" and qualified researcher. At the time such researchers were typically schooled in qualitative research strategies (such as action research or content analysis) taken from the social sciences and applied to the phenomena of performance. Snug within prevailing notions of research, researchers would then represent their findings in discursive prose and seek publication in reputable theatre journals.

However, in 1992 Alison Richards, writing on behalf of the ADSA Research Subcommittee, boldly claimed:

> We simply deny the legitimacy of the assumption that knowledge about performance, or the knowledges which come into being through performance, can adequately be represented within written or logico-verbal systems at all.[5]

Further, "research could be done through, or to steal Schechner's phrase *by means of* performance"[6] which can "validly be considered to be one of the communicative means by which it is proper to present research in demonstration/as publication."[7]

These were bold claims in 1992 and presented significant challenges for the newly formed higher education sector and its enthusiasm for research. Richards and her colleagues clearly understood that the gatekeepers of traditional research entities such as funding bodies relied on well-established and disciplined research methodologies from the quantitative and qualitative schools and would not be swayed by assertion alone. They reminded their colleagues:

> If we are going to argue that performance is an appropriate mode and method of research, we must be able to articulate why it is better for our purposes than other available or existing academic practices. What can

performance do better? What can be demonstrated through performance that is unavailable, clumsy or less clear through any other means?[8]

However, they were quick to dismiss the notion that all performance staged by academics in theatre and drama should now be simply reclassified as "research"; the field needed to move forward thoughtfully and with "rigour and restraint."[9] The credibility of performance as research had to be established – assertion alone would not bring it into being.

There were fundamental epistemological and methodological issues which needed to be addressed, and through a speculative list of field-defining questions, Richards raised key issues for performance as research around "intention, form and style, methodology, standards of execution and peer review procedures."[10]

Performance as research was developed with a firm eye on the protocols shared by all researchers from the sciences, humanities, and social sciences. This required sharp insight based on sustained reflection on both performance and widely accepted research protocols. There is no suggestion that performance as research could be used naively – as a term which would allow practitioner/academics to claim a *sui generis* research entitlement. From the beginning, ADSA recognized that for academics to become "performance as researchers," it was likely that significant changes to their performance practice would need to be made.[11]

In a series of papers, ADSA elaborated upon pivotal issues. Particular attention was paid to the nature and particularities of the research question or questions to be found in a study using performance as research. Connections were drawn between the refereeing process used to ensure quality in research publications and how a peer reviewing process may work for assessing quality in creative works produced through performance as research methodologies. A discussion paper presented by Richards drew together the various arguments in order to stabilize the field.[12]

Fifteen years on it is hard not to be impressed with the documents produced by ADSA. What is most significant is that these documents are written with two fundamental assumptions in mind: (1) performance can be understood as research with its own distinctive methodological inflections and epistemological frameworks and these had to be identified and argued, and 2) for performance as research to be established as a "legitimate" research strategy in its own right, its proponents needed to argue its case with an understanding of the established credibility tests and quality assurance processes that the research industry brings to bear on its whole operation – in the natural sciences, the life sciences, the social sciences, and the humanities. These are not documents of a "special interest" group pleading for "special consideration" so that performance as research can be understood and valued. Instead, they confidently claimed that performance as research can be a bold and credible new research strategy and that its case can be argued around fundamental and shared principles which all researchers can understand: the importance of question formulation, of methodological discipline, and peer review. Surely the future of performance as research in Australia would be secure and buoyant?

Missed opportunities

In 1996 a grant was awarded for a project to undertake "a comprehensive study of research outputs in art, craft, design, music and drama in order to develop a set of indicators and weightings in the creative arts."[13] This *Research in the Creative Arts Project* was led by Dennis Strand and has become popularly known as the Strand Report. The research project was a direct response to attempts by the Australian Government to assess the quality of research outputs by counting research publications and other bibliometric measures to ensure comparative analysis and rank. In 1994 and 1995 creative research outputs were audited in two categories, categories H (Designs) and J (Creative Works), which sought to include the non-traditional research outputs from researchers in the creative arts, media, and design. The Strand Report,[14] delivered in May 1998, promised to lay out both a rationale and means of operationalizing a system of indicators and measures which would enable creative artists and performance researchers to build a research profile and increase research funding to their disciplines.

Regrettably the systemic impact of the Research in the Creative Arts Project was minimal. Apart from another brief flirtation with the H and J categories in 2002, the research outputs of Australia's researchers in the Creative Arts went largely uncounted and they remained on the periphery of Australia's main research game. There are a number of reasons for this lack of impact, but perhaps one of the most influential was the decision of the project's advisory group to argue a case for "the notion of *research equivalence* as an appropriate and valid concept for recognition of research-based practice and performance in the creative arts."[15] Even at the time the report noted the term "research equivalence" might be misconstrued to denote activities that are "not quite research," or are "second-grade research."[16]

It would seem that the possible misconstruing Strand feared came to pass. As an idea, research equivalence failed to gain traction across the higher education sector perhaps because it held within its own internal logic the proposition that creative practice (including performance) was not inherently "research." Instead, it was cast as something "other" which, only in an act of special pleading, could be considered to be the equivalent of research.

The report asserted that "the outcomes of creative arts research will be different from, but equivalent to, research in other disciplines, removing the necessity of having to artificially 'shoehorn' some kinds of creative arts research into a traditional research model."[17] This is telling for it suggests that the work undertaken by ADSA in the early 1990s was not embraced by the Strand Project. Certainly the work by the ADSA researchers argued for the distinctiveness of performance as research but sought to meet the demands and protocols of traditional research in an engaged and productive way. There is no sense that the ADSA authors were docilely 'shoehorning' what they held dear about performance as research into an unsympathetic and hegemonic model of traditional research.

Since the Strand Report, performance as research has struggled to secure proper recognition. One attempt to advance the project occurred in 2001, when the Australian Research Council funded a special project called "Towards a research strategy for the creative arts: Creative practice, publication and research training," which was facilitated by the Australian Academy of the Humanities and the Australian National Coalition for the Creative Arts. The project involved three national symposiums that aimed to "debate the past, present, and future for innovation within the creative and design disciplines," and resulted in the edited publication *Innovation in Australian Arts, Media and Design: Fresh Challenges for the Tertiary Sector.*[18]

It is beyond the scope of this chapter to detail the history of the various research assessment schemes since then which have moved across the research policy stage and have implications for performance as research and creative arts research generally. Australian academics have lived through the period of the Research Quality Framework (RQF) from 2005–7 which aspired to measure both the quality and impact of research, and with the election of the Rudd government in 2007, a new policy framework called Excellence in Research (ERA) is being developed.

Challenges going forward

While performance as research has struggled to find its place in the larger research world of competitive grants and publications, it has prospered in the area of postgraduate research. In recent years an increasing number of candidates chose to use performance as research as the preferred methodology for their PhD or Masters by research. As a result this research approach is now thriving and well installed in a number of leading Australian universities.[19] This has challenged many of the positivist research methods courses offered by universities and problematized much common wisdom about postgraduate research training, research design, research outputs, supervision, and examination. In a practical sense, performance as research is now being advanced not through a national policy strategy, but through the incubator of higher research degree study.

The final section of this chapter will discuss three matters currently shaping Australian understandings of performance as research.[20] The first two build on matters signaled in the preliminary reflections and discussion papers published by ADSA in the early 1990s, namely the "problem" of the research problem, and secondly, how we may make creative research outputs, including research performances, available for sustained and verifiable peer review. The third matter to be discussed is methodological and suggests that the future of performance as research may lie best in aligning it with a more widely applicable research strategy known as practice-led research.

1 The 'problem' of the research problem

One essential protocol followed by the researchers working within the traditional paradigms of qualitative and quantitative research is that very early in a project they must clearly set out a research question which will lead the study. This is an essential requirement for all convincing research and as a result, research grants require a clear and upfront statement of the problem and PhD candidates in Australian universities are required to state their research question often within months of beginning their long and strenuous research degree. As a matter of course, applicants are asked to give a clear statement of the problem; to set out aims and objectives and the research questions to be answered; and to list the hypotheses to be tested. Statements of purpose, background, relevant literature, significance of the research problem, and definitions of key terms follow.

However, performance as research departs from this traditional approach. Many performance practitioners do not approach their creative research with a problem. Indeed, they may be led more by what can be described as "an enthusiasm of practice": something which is exciting, something which may be unruly, or indeed something which may be just becoming possible as new technologies, performance sites, or theoretical perspectives permit. Frequently, they prefer to construct experiential starting points from which further practice follows; they begin practicing to see what emerges and this research trajectory is likely to be individualistic and idiosyncratic. This is not to say these performance practitioner/researchers do not operate without larger agendas or emancipatory aspirations, but they are seldom assisted by setting sharply specified problems at the outset of a project. Certainly the research question or problem will be able to be clearly specified at some point in the project, but with performance as research it must be recognized that problem definition may be unstable for as long as practice is ongoing.

The ADSA discussion papers circle around this question of the research problem, sensing its importance. As we have become clearer about the nature of the research problem in performance as research, we can identify the fundamental difference of approach between the performance practitioner/researcher and the more traditional researcher following established quantitative and qualitative research approaches.

2 Ensuring research performances are available for sustained and verifiable peer review

Another area of challenge to traditional research protocols lies in the nature of the research outputs which flow from the performance as research process.

Practitioner/researchers insist that their research outputs and claims to knowledge must necessarily be reported through the symbolic language and forms *specific* to performance. This challenges traditional ways of representing research findings. Practitioner researchers across all of the creative arts believe it is folly to seek to "translate" the findings and understandings of performance into the numbers (quantitative) and words (qualitative) preferred by traditional research paradigms. They argue that to report research primarily in words or numbers can only result in the dilution and ultimately the impoverishment of the epistemological content embedded and embodied in performance. Consequently, the performance/researcher asserts the primacy of performance as a research output and join their colleagues in the creative arts by acknowledging that for the choreographer it is the dance, for the designer it is material forms, for the poet it is the sonnet, and for the 3-D interaction designer it is the computer code and the experience of playing the game which stands as the research outcome.

These "symbolic orders of knowing" conform to established definitions of research. The OECD definition, which is quoted in all of the documentation from Australian's Department of Education, Science and Training, states:

> Research and experimental development (R&D) comprise creative work undertaken on a systematic basis in order to increase the stock of knowledge, including knowledge of man, culture and society, and the use of this stock of knowledge to devise new applications.[21]

In the United Kingdom, the definition of research is even more tightly aligned with the alternative research outputs from practice-led researchers, where "research" for the purpose of the 2008 RAE is to be understood as:

> Original investigation undertaken in order to gain knowledge and understanding. It includes [...] the invention and generation of ideas, images, performances, artefacts including design, where these lead to new or substantially improved insights and the use of existing knowledge in experimental development to produce new or substantially improved materials, devices, products and processes, including design and construction.[22]

While there is international recognition that "legitimate" research outputs extend beyond the words and text preferred by traditional research paradigms, there is one further important point to be made.

A key credibility test for all research is that outputs are available for effective and sustained peer review. For traditional research, text-based publication is the spine of the system. Once in print, research findings, conclusions, and truth claims move effortlessly around the world for review by peers. The authority of the whole research industry is built around publications, peer assessment, and citation.

What does this requirement for sustained peer review mean for practitioners following a performance as research methodology? Does the time-based nature of the performing arts mean that peer review can only be applied while the performance is being staged and, if so, how can performance outputs in theatre, drama, music, and dance ever be made available in a sustained way for ongoing review? Must it necessarily be that when the performance is over, so are the possibilities for peer review? If so, can performance meet the requirements of the research industry?

Australian researchers including those from the Australasian Centre for Interaction Design (ACID)[23] are currently addressing this by building a citational infrastructure for performance. This is a digital platform for commenting on and annotating works and allowing peer reviewers and artist/researchers to interact and exchange views. The performance, whether of theatre, music, or dance, is recorded and replayed digitally and can be manipulated to allow annotation and comment to be made at specific points in the work. Such a system does not completely overcome the need for peer reviewer presence at the live performance, but it does go some way towards enabling peer review and publishing to be undertaken with an experience of the work present.

3 Performance as research – part of a larger research strategy?

In the early 1990s performance as research seemed an appropriate and distinctive research strategy to meet the needs of theatre, dance, and music researchers. Since then, as the number of projects has increased and higher research degree students have used performance as research more frequently, two methodological tensions have arisen. Firstly, it was discovered that performance as research could be informed by, and draw heavily on, certain research strategies and methods from the qualitative tradition and it seemed reasonable to borrow from, adopt, and modify a cluster of research strategies around reflective practice, reflection in action, action research, and grounded theory.

However, just as performance as research was incorporating selected qualitative research strategies and methods it also began connecting with a range of alternative strategies being developed in other creative arts disciplines. A host of terms have sprung up to name these practitioner-based approaches to research in the creative arts and design. They include creative practice as research, practice-based research, studio research, practice as research in performance, arts informed research, and practice-led research. In recent years practice-led research has become a prominent term for effectively describing a research approach that enables practitioners to initiate and then pursue their research through practice. In 1996 the visual artist and academic Carole Gray was able to confidently propose:

> By practice-led I mean firstly research which is initiated in practice, where questions, problems, challenges are identified and formed by the needs

> of practice and practitioners; and secondly, that the research strategy is carried out through practice, using predominantly methodologies and specific methods familiar to us as practitioners.[24]

It is important to note that Gray's definition refers to the difficulties noted earlier around the distinctively emergent nature of research questions with this approach, but goes on to suggest that creative practitioners can turn to the methods of their practice and re-purpose them to create an arsenal of research methods which serve their research needs. It is these re-purposed methods of practice which help practice-led researchers (including those from performance as research) build a systematic methodology to counter the view that all "practice is research, undisciplined, [and] without knowledge of what research paradigms are."[25]

What may be the advantages to consolidating this myriad of approaches and nomenclatures into a single strategy such as practice-led research? To start, practice-led research best describes what practitioner-researchers actually do. It captures the nuances and subtleties of their research process and accurately represents that process to research funding bodies. Above all, it asserts the primacy of practice and insists that because creative practice is both ongoing and persistent, practitioner researchers do not merely "think" their way through or out of a problem, but rather they "practice" to a resolution.

Additionally, the term "practice-led research" describes a research strategy which is relevant across the arts and design – a term which consolidates the field and adds the methodological scale and scope likely to strengthen the presence and impact of research from and by artists, designers, or creative practitioners. The field of performance as research is considerably enlarged when it is aligned with practice-led research and the wide range of disciplines it can effectively serve. Perhaps there is an important lesson to be learned from the outcome of the Strand Report. Even though drama academics had begun the rigorous thinking to develop performance as research, the wider research community in the creative arts did not recognize the potential it offered. Perhaps it was understood too narrowly, thought to be of interest to performance researchers only, and so the field lurched down the somewhat unproductive path of "research equivalence."

Conclusion

There is still much to do to consolidate and stabilize the term practice-led research and many of the questions raised by ADSA 20 years ago still need to be pursued.[26] While Gray offers us a useful beginning definition for practice-led research, its subtleties and contours need more careful mapping. Recently, my colleague Daniel Mafe and I elaborated upon the definition of practice-led research this way:

> It is a research strategy specifically designed to investigate the contingencies of practice by seeking to discipline, throughout the duration of the

> study, the ongoing emergence of problem formulation, methods selection, professional and critical contexts, expressive forms of knowledge representation and finally the benefit of the research to stakeholders.[27]

If it is possible to incorporate practitioners of performance as research into the workings of practice-led research, then the potential is likely to reach further than creative arts research only. Practice-led research has the potential to become a potent research methodology in interdisciplinary fields as diverse as arts and cultural innovation, online education, virtual heritage, creative retail, cultural tourism, and business-to-consumer interaction. In many ways the time is yet to come for performance researchers and their interdisciplinary colleagues who choose to collaborate around a practice-led research strategy, for such an approach will enrich every team seeking to address the significant research problems which confront us – serious problems from national security to community cohesion and social innovation. When this happens, performance practitioners will be transcending the important research needs of their own community and contributing their creative and research skills to benefit national and international research agendas.

Notes

1. ADSA is the Australasian Association for Theatre, Drama and Performance Studies. Formed in 1977, the original name was the Australasian Drama Studies Association. The association changed its name in 2003 to reflect the diversity of research and performance interests of the membership. However the acronym has been retained. For further information, *see* http://www.adsa.edu.au/
2. Members of the Performance as Research Subcommittee in 1995 included Alison Richards, Bill Dunstone, Gordon Beattie, and Tom Burvill.
3. Alison Richards, "Appendix A: Performance as Research/Performance as Publication," *see* http://www.adsa.edu.au/research/performance-as-research/appendix-a.
4. Alison Richards and B. Dunstone, "Appendix B: Some Issues in Performance as Research," *see* http://www.adsa.edu.au/research/performance-as-research/appendix-b.
5. Richards, "Appendix A: Performance as Research/ Performance as Publication."
6. Ibid.
7. Richards and Dunstone, "Appendix B: Some Issues in Performance as Research."
8. Richards, "Appendix A: Performance as Research/ Performance as Publication."
9. Ibid.
10. Ibid.
11. See ibid.
12. Alison Richards, "Discussion Paper: Performance as Research/Research by Means of Performance," *see* http://www.adsa.edu.au/research/performance-as-research/performance-as- research. Presented at the Australasian Drama Studies Association Annual Conference, Armidale NSW July 1995 on behalf of the Performance as Research Subcommittee.
13. The *Research in the Creative Arts Project* was funded by the Australian Government through the Evaluations and Investigations Program, the Higher Education

Division, and the Department of Employment, Education, Training and Youth Affairs.

14. Dennis Strand, "Research in the Creative Arts: Executive Summary and Recommendations," Canberra School of Art, The Australian National University, http://www.dest.gov.au/archive/highered/eippubs/eip98-6/execsum.htm.
15. Ibid.
16. Ibid.
17. Ibid.
18. R. Wissler et al., eds., *Innovation in Australian Arts, Media and Design: Fresh Challenges for the Tertiary Sector* (Flaxton: Post Pressed, 2004).
19. Some Australian Universities in which postgraduate students can follow practice as research include: Queensland College of Art (QCA); Griffith University; University of Technology Sydney (UTS); Sydney College of Arts; the University of Sydney; College of Fine Arts (COFA), University of New South Wales; Victorian College of the Arts (VCA), University of Melbourne; RMIT University; Victoria University; Monash University; Swinburne University of Technology; the University of Adelaide; the South Australian School of Art; Edith Cowan University; and Queensland University of Technology (QUT).
20. The three matters for discussion have been selected by the author and are not the result of an empirical survey of practitioners in the field.
21. OECD, *Frascati Manual*, Fifth edn (2002), 30.
22. RAE, "Definition of Research and Eligible Outputs 2008," *see* http://www.rca.ac.uk/pages/research/definitions_of_research_3091.html.
23. ACID, "Designing the User Experience," ACID Press, *see* http://www. interaction-design.com.au/index.php?option=com_content&task=view&id=21&Itemid=72
24. C. Gray, "Inquiry through Practice: Developing Appropriate Research Strategies," *see* http://www2.rgu.ac.uk/criad/cgpapers/ngnm/ngnm.html.
25. R. Stewart, "Practice vs Praxis: Constructing Models for Practitioner-Based Research," *TEXT* 5, no. 2 (2001): 4.
26. *See also* Brad Haseman, "A Manifesto for Performative Research," *Media International Australia Inforporating Culture and Policy*, no. 118 (2006); ibid., "Rupture and Recognition: Identifying the Performative Research Paradigm," in *Practice as Research*, ed. E. Barrett and B. Bolt (London: I. B. Tauris, 2007); ibid., "Tightrope Writing: Creative Writing Programs in the RQF Environment," *TEXT* 11, no. 1 (2007); L. Duxbury, E. M. Grierson, and D. Waite, eds, *Thinking through Practice: Art as Research in the Academy*, 2nd edn (Melbourne: The School of Art, RMIT University, 2008).
27. Brad Haseman and D. Mafe, "Acquiring Know-How: Research Training for Practice-Led Researchers," in *Practice-Led Research/Research-Led Practice in the Creative Arts*, ed. Dean and Smith (Edinburgh: Edinburgh University Press, 2009).

7
Locating the Artist-Researcher: Shifting Sites of Performance as Research (PAR) in Canada

Laura Levin

A compelling case has been made in the last decade for approaching performance as a viable form of critical inquiry and research. Propelled by new granting schemes and evolving disciplinary alliances, artists in Canada have engaged in dynamic experiments in performance praxis, the embodied exploration of theoretical concepts and research questions through the medium of performance. Given its location within the North American academic landscape, it would be reasonable to assume that the "performance as research" practices in Canada bear a close resemblance to those found in the field of performance studies in the United States.[1] While Canadian scholars in several disciplinary contexts have adopted performance studies paradigms, incorporating practice-based approaches in their research and teaching, the precise terms of this exchange differ significantly in the Canadian context, reflecting less the idioms of performance studies found in the United States and more the discourses and preoccupations of Canada's funding bodies and cultural institutions.

Rather than offering a comprehensive survey of PAR projects in Canada,[2] which is outside of the scope of this chapter, I hope to illustrate how specific material conditions within Canada's academic and artistic environments have shaped approaches to performance as a research activity. To do so, I will briefly highlight three areas in which performance is being explored as a site of knowledge production, focusing primarily on performance in Ontario as an illuminating case study. These three related axes, which I will refer to as research/creation, urban intervention, and pure research, provide a complex matrix for defining the "artist-researcher" in Canada. This figure has become an important commodity within an environment of global economic competition, one inextricably linked to national and civic projects of self-identification. I will also reflect on the competing understandings of arts-based research that coexist within the Canadian performance studies setting. In sorting through discontinuous and often strategic definitions, it becomes clear that the value of practice as research can be found not only in the dynamic performances produced within its frame but also, and often more importantly, in the institutional prerogatives that this experimentation puts into relief.

Research/creation

One of the most important models of performance as research in Canada is "research/creation." This term has been popularized by Canada's federal funding agency, the Social Sciences and Humanities Research Council (SSHRC). In Fall 2003, SSHRC launched the Research/Creation in Fine Arts Pilot Program, which provided large multi-year grants to university faculty (usually in amounts over $100,000) for the development of arts-based projects with a significant research component. According to SSHRC, the program responds to "a new 'practice-based' or 'performance-led' research paradigm" that is emerging among artist-scholars within the academy.[3]

From 2003–8, the research/creation program has been at the centre of practice-based scholarship in the fine arts in Canada, funding a wide range of research projects that employ or lead to performance creation. A number of the projects involve historical re-enactment, including, for example, *Shakespeare and the Queen's Men* (Alexandra Johnston), an attempt to "recreate the staging conditions of a sixteenth-century touring company in order to study and test scholarly theories about acting styles and repertory through performance practice"[4] and *The Zoo Project* (Darcey Callison), an exploration through choreography of past representations of masculinity in popular dance. Other projects foster new forms of performance – for example, *Networks Performance* (John Celona), a project designed to create multimedia performances using a broadband network, and *Sense-stage* (Chris Salterone), which aims to "develop new wireless sensor and software tools" for "interactive real time performance."[5]

For artists at Canadian universities, practice as research has become their bread and butter, the precondition for securing the funding and attention their work deserves. This development is influenced not only by the fact that the Canada Council for the Arts, the major arts grant agency, is directing university artists to SSHRC to fund their practice,[6] but also by growing demands that they contribute to university research profiles, which come in the form of changing tenure standards, submission to research ethics review, and the adoption of new kinds of professional self-description. Many of the newly confirmed artist-researchers embrace this shift, seeing it as a recognition of intellectual rigor in the arts and as a challenge to binaries in the academy (intellectual/technician, theory/practice, and so on).

Still, it is important to take a step back to consider the precise understanding of research that is being promoted by "research/creation" and the impact that it may have on the emerging phenomenon of performance as research in Canada. First, there is the issue of how research is conceptualized within this rubric. SSHRC guidelines state that the goal of the research/creation program is to fund projects that are "scholarly" in nature. Definitions of research/creation stress that artistic practice be approached through the languages of the social sciences and humanities, SSHRC's main constituencies. SSHRC defines Research/Creation as "sustained, reflective research (with a fully developed scholarly apparatus) set directly and actively within the creative or artistic process itself."[7] Research/Creation projects must "address clear research questions, offer theoretical contextualization within the relevant field or fields of literary/artistic inquiry, and present a well-considered methodological approach."[8]

These definitions recently sparked debate at a program information session at the Ontario College of Art and Design. According to presenter Geoffrey Rockwell, several artists commented that SSHRC was sending the message that it is funding "research/creation (not art)"[9] and were unsure about how their practice would fit into this program. While these comments might be read as reinstating researcher/artist binaries, I think they also point to potential shortcomings of the SSHRC model, perhaps signaled in the slash between research and creation. Research/creation seems to imply research, as it is traditionally conceived, that takes place alongside the artistic process and generates an artwork. SSHRC has advised research officers that successful projects will disseminate research results *both* in the form of an artistic production *and* scholarly publications. In this context, it is easy to read research/creation as research *and* creation, rather than a hybrid third term that merges the two. This is exacerbated by the fact that nowhere in the SSHRC guidelines is there formal recognition that the art process *is* the method of research and, as such, may require a different set of evaluative criteria. Thus, the term research/creation is in danger of upholding the very binaries that it purports to undo; the slash between research and creation begins to look like an either/or.

Adding to these tensions around research/creation is the problem of output. While the PAR tradition in performance studies emphasizes process over product, research/creation projects must "meet peer standards of excellence and [be] suitable for publication, public performance or viewing."[10] The focus on successful product reflects SSHRC's mandate to publicly demonstrate the impact of research through concrete deliverables. This has become critical with the growing push by the federal government to represent Canada as a major actor on the global stage. In its recent publication, SSHRC defines its role as meeting the needs of "a competitive, global, knowledge-based society and economy."[11] Establishing Canada's "world-class excellence"[12] is essential to an image of nation that can be marketed worldwide, helping it advance in the global competition for "ideas and talent,"[13] stimulate investment, and develop technological advantage. Within this knowledge-based economy, where creativity is part of the country's global marketing scheme, the creations of artist-researchers have become valuable commodities. It is not surprising to see that most funded research/creation projects are experiments with new technologies, coinciding with policy directives to "mobilize science and technology to Canada's advantage."[14]

The Research/Creation program, now at the end of its pilot term, has been suspended pending agency review. There has been much speculation about the future of the program. Some see it turning its focus toward collaborations between artists and scientists, which would draw upon the arts in a more explicit way to serve technological ends. Others predict that the responsibility for funding university artists will be shifted to the Canada Council. The input of artist-researchers will be essential to defining the future of this PAR model in Canada.

Urban intervention

While research/creation focuses on the university as performance lab, making the training of students one of its main objectives, a second model of performance as research – urban intervention – finds its laboratory in the spaces of everyday life. Unlike research/creation, which is designed to advance knowledge in the fine arts, urban intervention takes its cue from a much broader range of disciplines, including environmental studies, cultural studies, and sociology. Here, the experience of urban life is the central object of study, with a particular focus on the ways in which bodies shape and are shaped by the environments that they inhabit. The terms "urban intervention" and "site-specific" have become virtual stand-ins for performance studies in Canada, as a significant number of Canadian performance theorists have concentrated in this area.

Rose Bianchini and Christian Smith offer an accessible definition of urban intervention. "Often artistic and always political," they write, "the modern practice of urban intervention carries on the avant-garde project of attempting to unite art, politics and everyday life, actively challenging conventional ways of seeing, inhabiting, moving through, and engaging with the city." These interventions are both "performative – they encourage the adoption of identities and modes of being that are different from the everyday," and "participatory – they actively seek to inspire and incite interactivity, problematizing the role of the passive spectator."[15]

Urban intervention, particularly prominent in Toronto, has functioned both as a form of civic boosterism, an attempt to construct a new conception of the city, and as a response to the twinned influences of privatization and corporatization, threatening access to public space and leading to the spatial segmentation of urban life. Groups like the Toronto Public Space Committee have kept alive Guy Debord's critique of the "society of the spectacle," mobilizing against the replacement of public space with spaces of consumption. TPSC campaigns include the De-fence Project, where participants remove fences around residents' private property at their request, and Art Attacks, where art is installed over top of commercial advertising spaces.

Toronto's urban interventionists have turned to performance, and to embodied "practice" in the tradition of Michel de Certeau, as a testing ground for alternative uses of space and modes of social relation. The Toronto Psychogeography Society (TPS) takes as their point of departure the experimental research traditions of the Situationists, who set out to "study the specific effects of the geographical environment, consciously organized or not, on the emotions and behaviour of individuals."[16] Members of TPS meet every Thursday night to engage in exploratory walks, which allow participants to "step out of their daily routine and explore the city's overlooked corners to imagine the dynamics of a better future urban environment."[17]

This kind of performative urban ethnography is also present in the work of groups like [murmur]. [murmur] is an oral history project that documents stories about sites in Toronto and makes them available through a number that can be dialed via cell phone at those locations. Residents of Toronto, whose voices animate the audio-tour, produce an intimate mapping of the city, an urban knowledge based in memory that troubles official discourses about the city's meanings. As Chris Eaket notes, [murmur]'s spectator is "active in the *performative production of space*," engaging in speech acts "that performatively produce the built environment as it is experienced."[18]

Mammalian Diving Reflex (MDR) takes this kind of research in a more explicitly theatrical direction. Composed of a group of "research artists," the company has engaged in a form of experimentation that they call "social acupuncture" – an attempt to break "system-wide holding patterns" that compromise the healthy functioning of the social sphere.[19] Social acupuncture challenges the "lack of free public space for unstructured discourse,"[20] which produces feelings of social detachment in urban life. In *The Talking Creature*, participants recruit strangers from a neighborhood and gather later to converse informally with their found subjects. The experiment serves as a form of inquiry into the lives of those community members put on display and also forces spectators to move outside of their private shells, to experience "unfettered and fearless conversation."[21] This research is extended in the work, *Diplomatic Immunities*, where company members roam the city interviewing residents and then present their stories later that evening in an interactive show that is part ethnography and part confessional. In their words, "Diplomatic Immunities takes the body of the city, throws it on stage, cuts it open and holds the contents up to the light."[22]

While research/creation and urban intervention advance different models of performance as research, they face similar challenges as both are hailed by the rhetoric of global competition. Toronto, like other North American cities, has turned to the arts to help position itself as a global cultural centre. This project of self-branding, outlined in Toronto's ten-year culture plan (developed in 2003), adopts the civic model put forward by urban gurus like Richard Florida, which see the "knowledge worker"[23] – artists scientists, and so on – as holding the key to an expanding economy, using their bodies, creativity, and lifestyles to sell Toronto to its citizens and to the world.

Despite the anti-commodification stance of many in the urban intervention movement, Toronto's artist-researchers and the official agents of creative capitalism often find themselves sharing the same urban development script.[24] Not only do the interventionist performances become ideal advertisements for the cultural vibrancy of the city – especially when spectacle eclipses political content – but also artists can support the neoliberal agenda of interurban competition in their enthusiastic attempts to drum up civic pride. As the pressure to adopt Toronto's Creative City vision grows, urban artists are learning how to negotiate these new cultural frameworks without allowing them to dilute the radical social critique to which urban intervention aspires.

Alternative directions: Pure Research

It would be naive to imagine that any form of performance research could operate entirely beyond the global circulation of capital and the state-funding ecosystem that sustains the arts in Canada. By becoming aware of the material structures that shape this kind of work, we are more likely to seek alternative ways of defining and conducting research, ones that might allow critical inquiry and experimentation to stand as the *ends* of performance.

Artist-researchers interested in this more open approach might take their cue from the methods pioneered by the dramaturgical theatre company, Nightswimming. Influenced by the vibrant history of devised theatre and collective creation in Canada,[25] which has traditionally placed creative process at the centre of art making, Nightswimming runs a unique program called Pure Research, which offers artists space and funding to pursue "a provocative theatrical question."[26] Artists working with Nightswimming have explored questions like, "What happens when you place the audience and/or a performer "inside an 8′ × 8′ box,"[27] and, "What are the complex emotional and communication issues that accompany the transition from speaking to singing and back?"[28] Not surprisingly, Mammalian Diving Reflex's first experiments with social acupuncture took place in the early stages of the Pure Research program.

Nightswimming provides a lab context that is free from the pressures of creating a polished work; no public performance is required. The program description offered by artistic director Brian Quirt implicitly troubles the consumable "Creation" side of Research/Creation: "Nightswimming's Pure Research program is not a new model for creation but suggests a dramaturgical approach that escapes the inherent restrictions that any goal-oriented model imposes."[29] This approach, which accepts failure as a potential and acceptable result of research, not only breaks with the excellence-oriented frame of Research/Creation, but also gives professionals a place to reflect on the theatrical process in a setting removed from industry pressures to put bodies in seats.

Pure research offers a glimpse of an alternative way forward, a model of performance as research that should be nurtured alongside emerging institutional paradigms. Given that pure research grows out of a utopian impulse to create spaces for thinking about art on its own terms – resisting the society of the spectacle and the university assessment microscope – there is a tremendous opportunity to use this and similar process-oriented paradigms not only to advance knowledge about form (as the word "pure" tends to imply), but also to create dissident knowledges about the social. In so doing, this type of research can provide a forum for questioning power structures and addressing pressing issues within diverse communities.

Notes

1. Only one program conferring MA/PhDs in Performance Studies has been created to date (University of Calgary). Other institutions like York University, University of Guelph, and University of Toronto offer graduate courses in this area.
2. There are other models of performance as research in Canada that deserve attention. These can be found in a variety of areas: devised theatre and collective creation, performance art, research-based performance in health studies, and drama in education, to name a few.
3. SSHRC, "Congress," Social Sciences and Humanities Research Council, *see* http://www.fedcan.ca/congress2005/programs/6_1.htm.
4. "Shakespeare and the Queen's Men Conference: Call for Papers 2006," Records of Early English Drama (REED) Centre for Research in Early English Drama, *see* http://www.reed.utoronto.ca/QueensMen/abstract.html.
5. CAFAD, "Cafad Newsletter," The Canadian Association of Fine Arts Deans, *see* http://www.cafad.com/cafnews%20jun07.pdf.
6. Geoffrey Rockwell, "Research/Creation, Again," geoffreyrockwell.com: http://www.philosophi.ca/theoreti/wp-content/uploads/notes/001092.html.
7. SSHRC, "Congress."
8. Ibid., "Definitions," Social Sciences and Humanities Research Council, *see* http://www.sshrc.ca/web/apply/background/definitions_e.asp.
9. Rockwell, "Research/Creation, Again."
10. SSHRC, "Definitions."
11. Ibid., "Framing Our Direction," Social Sciences and Humanities Research Council, *see* http://www.sshrc.ca/web/about/publications/framing_our_direction_e.pdf.
12. Ibid.
13. Ibid.
14. Ibid.
15. Rose Bianchini and Christopher Smith, "Crash Course in Critical/Creative Urban Intervention," Anarchist Free University, *see* http://anarchistu.org/cgi-bin/twiki/view/Anarchistu/CrashCourseInCritical.
16. Guy Debord, qtd in Simon Ford, *The Situationist International: A User's Guide* (London: Black Dog Publishing, 2005), 34.
17. TPS, "Who," Toronto Psychogeography Society, *see* http://psychogeography.ca/who.htm.

18. Chris Eaket, "Project [Murmur] and the Performativity of Space," *Theatre Research in Canada* 29, no. 1 (2008): 39–40.
19. Darren O'Donnell, *Social Acupuncture* (Toronto: Coach House Books, 2006), 48.
20. Ibid.
21. Ibid., 52.
22. MDR, "Diplomatic Immunities: The End," Mammalian Diving Reflex, *see* http://www.mammalian.ca/template.php?content=shows_diplomatic.
23. Richard Florida, *The Rise of the Creative Class* (New York: Basic Books, 2002), 9.
24. *See* Laura Levin and Kim Solga, "Building Utopia: Performance and the Fantasy of Urban Renewal in Contemporary Toronto," in *Canadian Association for Theatre Research Conference* (Vancouver: 2008).
25. *See* Ric Knowles and Skip Shand, eds, *Canadian Theatre Review (CTR): The Process Issue* 97 (Winter 1998); Bruce Barton, ed. *Collective Creation, Collaboration and Devising* (Toronto: Playwrights Canada Press, 2008).
26. Nightswimming, "Pure Research with Nightswimming," Nightswimming, *see* http://pages.interlog.com/~bquirt/research.html.
27. Camelia Koo, "The Box," Nightswimming, *see* http://pages.interlog.com/~bquirt/research_koo.html.
28. Guillaume Bernardi, "Exploring the Land between Speaking and Singing," Nightswimming, *see* http://pages.interlog.com/~bquirt/research_bernardi.html.
29. Brian Quirt and D. D. Kugler, "Editorial," *Canadian Theatre Review (Creative Research and New Play Development)* 119 (Summer 2004): 3.

8

Social Performance Studies (in China): Between the Real and the Virtual

William Huizhu Sun and Faye Chunfang Fei

In traditional theatre studies, the division between performance and research is quite clear. The former refers to the making of theatre; the latter refers to scholarly writings about the former, based on critical analysis, observation, and/or historical contextualization. Over the last 60 years or so, the subjects taken up by theatre researchers have grown from dramatic literature and history to encompass areas such as acting, directing, design, and theatre management – often referred to as the performance aspects of theatre. The term 'performance studies' is frequently used to indicate these new areas of inquiry.

Originally, the term performance studies (PS) was coined in the late 1970s by Richard Schechner, a New York University (NYU) drama professor and experimental theatre director, and Victor Turner, an anthropologist deeply interested in "social drama," to indicate a resolute break from traditional theatre studies. The year 1979 marked its first institutionalization, as NYU's Department of Graduate Drama officially changed its name to the Department of Performance Studies. This research field applies the model of theatre studies to subjects such as tribal ritual, street demonstration, happenings by musicians or visual artists, and so on. From its inception, performance studies began to dissolve the more traditional separation of practice and research – as Schechner notes:

> A number of performance studies scholars are also practicing artists working in the avant-garde, in community-based performance, and elsewhere; others have mastered a variety of non-Western and Western traditional forms. The relationship between studying performing and doing performance is integral.[1]

NYU's program was the first of its kind in the world. In 1985, another PS department was established at Northwestern University in Chicago. This program developed out of a Department of Communication (Speech and then Oral Interpretation). While the NYU model takes an anthropological approach, viewing all human performance activities as legitimate subjects of study, the Northwestern model studies performance from the perspective of oral performance, focusing on varying forms of speech communication in different social situations. Finally, theatre scholars in the United Kingdom also use the concept of PS – not to refer to the kinds of performance beyond the theatre as their American colleagues do – but to the performance aspects of theatre as outlined in the first paragraph above.

Unlike the three established types of PS in the West – which we divide as theatre PS (mostly in the UK), human PS (NYU model), and oral PS (Northwestern model) – "social performance studies" is unique, with Chinese characteristics. It is a new research field started at Shanghai Theatre Academy (STA) in collaboration with diverse institutions. Taking performance as research, it focuses on studying urban professional performances by teachers, doctors, lawyers, and government officials.[2] By examining and/or inventing various liminal genres between the extremely real and the extremely virtual, social performance studies explores areas of training for social performance and the bridging of social and artistic performances.

Locating social performance studies in China

Between the real and the virtual, there are countless liminal genres of performance and many examples that challenge a simple separation between the two. For example, in traditionally "real" genres we can see many non-real parts, such as actor performances in "documentary" films, fabricated plots in talk shows and reality TV, and story enactments in commercials. On the other hand, in traditionally fictional works we see many so-called real materials, such as court documents and personal letters in semi-fictional dramas, documentary footage in feature films, celebrities appearing as themselves in otherwise fictional plays, and so on. Jerry Seinfeld's nine-year-long sitcom series bears his real name, while he plays a seemingly real character in his same profession. Stan Lai made a year-long daily sitcom in Taipei, rehearsing every day with the latest real news that was broadcast in the evening a few hours later.

Today's societies are saturated by all kinds of artifacts. Imaginary artistic works often predate live events; hence, life can also imitate theatre/performance/image – challenging the too-tidy notion that art imitates life. In the 1992 Los Angeles riots, television crews rushed to downtown LA to shoot scenes for national and international broadcasting, unwittingly instigating many TV watchers in other cities to follow the lead of the LA rioters. The film *Mad City* (1997), starring Dustin Hoffman and John Travolta, echoes this phenomenon. Travolta plays a laid-off security guard at a small-town museum. Hoping to blackmail the museum director to rehire him, he threatens him with explosives that he never intends to use. Hoffman plays a TV reporter who learns about the explosives and sees a group of school children in the museum. He immediately uses his cell phone to inform his boss of the (imagined) situation. The whole world learns about the "terrorist incident" through the snowball of airwaves that pulse from that small-town TV station. Countless media people rush to the museum, making the entire town a "mad city." The media is so hungry for a dramatic conclusion that in the end, Travolta's character is cornered with no choice but to ignite the explosives. In another feature film, *Wag the Dog* (1997), a government hires a TV producer to fabricate a scene wherein an enemy opens fire. They broadcast it as news in order to convince the congress to declare war. This story is similar to the US government's tall tale of Saddam Hussein's weapons of mass destruction but takes one more step, turning an oral performance into a theatrical performance. Who knows whether this type of fictional performance will be or has been imitated by government officials?

In the continuum from the real to the virtual, "real" killing takes place on a battlefield, in terrorist bombings, and in crime scenes, whereas in virtual worlds, killing is performed in Internet games on screen. Addiction to Internet games has become a serious social problem because its virtual violence may be psychologically damaging to young people, yet it is free from legal prohibition. Compared to Internet games, Murder Mystery Game, a popular form in Chinese cities in recent years, is closer to reality because it is played by a group of real people: the players sit together, each given a piece of paper with identities such as Killer, Policeman, Commoner, and so on. Their actions, however, are limited to speech and facial expressions. Although the "Killer" and "Policeman" may sit next to each other, no dramatic action is performed.

A murder on stage appears closer to one committed in the real world; for example, actors playing Hamlet and Laertes wield realistic looking swords to fight in a way that looks real, even if they do not harm one another. By comparison, the fighting in Chinese opera is ostensibly stylized and does not suggest that anybody could be killed or hurt. Fighting in realistic drama is far more convincing and could possibly frighten first-time theatergoers – and fighting in film is often similar. Except in rare cases, such as when Bruce Lee's son was accidentally killed by a real bullet rather than a blank, all killings on stage and screen only appear real.

Between the staged and the real, there is perhaps another genre: the ancient Roman gladiators' fight. Although gladiators actually killed their rivals, it was not an act of war, revenge, or hatred. The killing was purely for the entertainment of the spectators and fundamentally different from killing in real life. Compared to the others mentioned above, however, the gladiators' is closest to the real.

Thus, there are various layers of performative activities between the most virtual Internet game and the most real war:

real killing: war, execution, murder	gladiators' killing	stage killing	murder mystery	virtual killing game

Between any two of these, one can find or invent more layers. For example, between war and the gladiators, there is public execution, which is by all means real killing but is also a show intended to exhibit power. To be sure war can become a public show as well – American military commanders are very good at making missile strikes a television attraction. When they bombed Afghanistan and Iraq, people across the world saw few real victims but many bright lights from flying missiles on their television screens – a weapon of psychological warfare. And between stage killing and the sit-down Murder Mystery Game, social performance studies students at STA are inventing a stand-up murder mystery game in which players play specific characters in concrete situations as if in a loosely organized theatre piece without a fixed ending.

As Chekhov says, people are not feuding with one another or hanging themselves every minute. Everyday life is usually not as dramatic as killing.[3] Nevertheless, many nonviolent activities can also be seen on the real–virtual coordinate in the same way. Take driving as an example – one of the most mundane activities shared by people of various classes and cultures in modern societies. While daily driving is the most real, student driving on the real road with an instructor or examiner sitting on the side could be considered virtual and performative – a demonstration of skill displayed for the instructor/examiner. Driving a Formula-1 racecar is even more ostensibly virtual and performative. Children's bumper cars are more like toy cars, and their driving more like a rehearsal for real driving when the children grow up. On the most virtual end of the coordinate, needless to say, is the programmed driving game performed on/in front of the television or computer screen.

daily driving	student driving	F-1 racecar	bumper car	driving game on screen

These coordinates suggest many possible ramifications, mainly in two areas: education and training that have traditionally been seen as "real," and arts and entertainment that have traditionally been seen as "virtual" – although this, too, has become blurred as virtual technologies are increasingly used in training. For example, virtual driving has been used recently in some driving schools in Shanghai.[4] The use of virtual driving cabins cuts short the learning time on the road, and saves 40 liters of gas per student. This model has of course been used in pilot training for decades, due to the high cost of training in a real plane.

This model can also be adopted in other professional fields. The FBI Academy employs professional actors to play suspects in realistic settings and is much more effective than having students take turns playing officers and criminals. Because they do not recognize the players, it makes the practice far more serious and convincing.[5] Applied theatre professors from Australia have also used similar methods to help train policemen. They set up a super-realistic site in a small town, where professional actors playing criminals are to be discerned and arrested by officers in training. Compared to regular pedagogical tools, these stuntmen-like performers are far more efficacious instruments to help students gain required skills.

Teaching at police academies in China is still largely classroom-centered. Recently a few leading police academies decided to stop admitting high school graduates and recruit only on-the-job policemen and those with a college education.[6] Classroom-centered teaching can hardly turn inexperienced high school graduates into competent police officers – but changing the makeup of the incoming students may not be enough to make training in police academies more effective. The skills police officers need are highly sophisticated and practical – they are not only the kind of technical skills that can be gained in a scientific laboratory. What is needed is a social laboratory made up of scenarios and performers, including students playing police and staff performers playing their hidden targets.

However, this kind of performative lab is quite costly. Therefore, before using this laboratory, students can prepare by first playing onscreen detective games to sharpen their perceptions, and then performing in interactive live plays or the "stand-up murder mystery game" that STA students are creating in order to gain experience and hone relevant skills. Those three steps fit exactly in the three performative coordinates between the "real" and "virtual" on the above diagrams. They offer students hands-on and teamwork experiences, both of which are absolutely necessary in professional training yet unavailable in the traditional pedagogical model of classroom lecture plus examination.

All these problems reflect a profound defect in the Chinese education system. By and large, there is no theatre, performance, or even speech class in general education; students are always taught to be good listeners and written test-takers. They often do not know how to properly present themselves in public, especially in front of a video camera; nervousness and overacting are common. While more and more people, including government officials, have seen these problems in China, it is still very hard to put theatre into the school curriculum, because the current curriculum in elementary and secondary schools, designed around college entrance exams, is already very full, and art courses are not part of the college entrance exam. In universities, on the other hand, curricula are strictly divided into specific majors, which are determined for most students even before their first day of class. Classes in theatre or any kind of performance are for major students in those areas only. The new theatre appreciation classes some universities have begun to offer for non-majors are usually play reading or video watching, plus lectures, with little room for students to perform. Moreover, there are few university professors that can teach performance, be it artistic or social.

For these reasons, STA advocates the use of social performance studies in various humanities and social science classes, including acting in plays relevant to students' future professions, such as plays about doctors for medical students, legal plays for law students. This new approach needs a great deal of experimentation and research in pilot projects in selected schools. At STA, a theatre conservatory, a new graduate major in Social Performance Studies, and an undergraduate major in Theatre Education have been established in recent years to train students in applying theatre skills in non-theatre fields. Research projects have included "Lawyers and their Stage/Screen Representations," "Doctors and their Stage/Screen Representations," and "Journalists and their Stage/Screen Representations" – all projects aim at applying the above virtual–real coordinate to practical training. In each case, video screenings of various working situations is the first step, followed by discussion and then by enactments of similar situations, often in the form of Augusto Boal's forum theatre in which any *spectactor* can stop the play and replace an actor to develop a different action. After these stages, a real situation internship with supervisors takes place, before sending the students to real jobs. What is planned further is to develop interactive electronic games based on relevant situations. Games could be more efficient than human enactment in experiencing and learning to handle various situations. Human enactment is ultimately indispensable as it gives performers a deeper, visceral, feel of the experience.

Other projects such as "Human Role-Playing Games and Vocational Training," "Theatrical Theories Applicable to Classrooms," and "Emotional and Cerebral: Two Paradigms of Experiential Training" explore various ways in which performance can be an integral, or primary, part of the pedagogy, in order to turn students into knowledgeable and practical professionals who can work with their brains, hearts, hands, and mouths.

Compared to the more established types of PS, social performance studies is a fledgling field where there is not yet a fully tested curriculum. It is largely a project-oriented field bridging the theatre academy with various non-theatre related institutions as collaborators. It is hoped that in the future such collaborations will include international partners such as those in the Department of Corporate Performance at NIDA (Australia's National Institute of Dramatic Arts), London's RADA Enterprises, and Australia's applied theatre programs.

Notes

1. Richard Schechner, *Performance Studies: An Introduction*, 2nd edn (New York and London: Routledge, 2006), 1–2.
2. Editors' Note: this approach to social performance studies has much in common with the applied rhetoric field of the 1970s to 1980s Polish School headed by Jerzy Axer in Warsaw.
3. In Yermiluofu, *On Chekhov's Works* (Beijing: Writers' Press, 1957), 129.
4. Wan Yujun, "Virtual Driving Cabins Show Up in Shanghai for the 1st Time," *Labor Daily*, 4 December 2007.
5. Emily Colborn-Roxworthy, "Role-Play Training at 'Violent Disneyland': The FBI Academy's Performance Paradigms," *TDR: The Drama Review* (Winter 2004).
6. "Police Academies to Stop Recruiting Regular Undergraduate Students," *Southern Urban Post*, 2 January 2008.

9

Artistic Research – from Apartness to the Umbrella Concept at the Theatre Academy, Finland

Annette Arlander

During the last 30 years, the development of practice-based research in the arts in Finland has occurred mainly within arts universities, and has followed slightly different strategies in each, due to the requirements of various art fields. This chapter concerns the Theatre Academy (TeaK), where I studied directing from 1977–81 and now work as Professor of Performance Art and Theory and as Head of the Department of Research.[1] To some extent, the institutional history of research at TeaK coincides with my own journey into artistic research – a trip from theatre through performance to visual art and from space through place to landscape.

At TeaK the development has been from experiments with reporting artistic practice, through a dichotomy between doctorates with scholarly or artistic emphasis into artistic research. Using the spatial analogy, research has evolved from: (1) clearing a space for research within a professionally oriented arts university; through (2) developing a place for research using qualitative, pedagogical or phenomenological orientation; into (3) a landscape of various approaches to knowledge production in performing arts under the umbrella concept of artistic research. The choice of terminology reflects the focus on artist-based rather than generally practice-based research.

Development from dichotomy to differences

The Theatre Academy was founded in 1979, and provides the highest education in theatre and dance in Finland. Since 1988, postgraduate students at TeaK have been able to undertake doctoral degrees with the possibility of attaining a Doctor of Arts either in Theatre and Drama or in Dance. At this moment, TeaK has "produced" 16 doctorates, and Acta Scenica, a TeaK publication series, has published dissertations and reports on theatre, dance, and performing arts.[2] There have been only four doctoral works with artistic emphasis, where performances are examined as part of the dissertation.[3]

Though the official doctoral title clearly states that this is a degree in arts, a dichotomy was established during the 1990s between doctoral works with artistic emphasis and those with scholarly/research emphasis. However, in 2007, TeaK chose to change its strategy. A Department of Research Development was founded, and a professorship of artistic research established, in order to develop a theoretical basis for artistic research that is specifically suited for performing arts. The first tasks have been to abolish the dichotomy between scholarly or artistic emphasis, make possible a broad spectrum of research approaches, and encourage practicing artists to conduct research. Here pedagogical research has served as an intermediary. Today all new doctoral works have a research emphasis, contain artistic or practical parts, and undergo conventional academic procedure. All research undertaken at TeaK is considered artistic research, as befits an arts university. However, many crucial questions remain open.

In the following I explore these past changes through my own experiences as a doctoral student at TeaK in the 1990s and use the publication series Acta Scenica as a point of reference.

Acta scenica 1 – *Knowledge is a Matter of Doing*

The first research publication, *Knowledge is a Matter of Doing,*[4] consists of the proceedings of the symposium Theatre and Dance Artists doing Research in Practice, which took place at TeaK in October 1994. It was the first of its kind in Scandinavia in which theatre professionals, teachers from institutions training practitioners, and scholars from universities participated. A key topic was research work done in theatre laboratories, outside established institutions.

For Grotowski, "[k]nowledge is a matter of doing."[5] This observation can be extended to the whole field: you either do artistic research, try out various approaches and give them a chance to develop, or you keep on discussing the problems to guarantee that the results are appropriate.[6]

At the time of the symposium in 1994 I was writing my MPhil report as a young theatre director trying to combine her artist's training and studies in theatre research into an articulation of her work in English:

> Firstly, the dichotomy between a so-called scientific and a so-called artistic approach was crystallised for me in this project as being the difference between two ideas: to study the *influence of space* on the performance (both in the making and in the experiencing of it), or to study the *use of space* in a performance (both in making and experiencing it). Since I consider myself to be an artist more than a scientist, even though I am more interested in finding out something I do not know than in communicating to others something I already know or, if you wish, in finding evidence for something I believe I know, I chose the latter: how to use space, and how to do it following the logic of the play, not any other systematisation. The

question was how to use both the fictive space in the text and the concrete performance space and its surroundings as a starting point for creating a third kind of space that is a 'performance world'.[7]

"Some Conversations XI – A Report on the Series of Performances Produced by the TEE-Project," was examined in 1995. The unpublished report suggests that space and place can be used as a means of expression in a theatre performance – not only to facilitate and influence the meeting of performers and spectators or the success of the performance event, but also as a central starting point for creating the performance composition.[8]

Reading this reminds me of a remark by Susan Melrose (quoted from wishful memory), as she visited TeaK in 2002: "An artist always does more than she knows, and knows more than she can say."

Acta scenica 2 – *Performance as Space*

The second research publication, *Esitys tilana (Performance as Space*: 1998), was my dissertation – the first with artistic emphasis completed at TeaK. There were four examiners participating as opponents at the public defence to guarantee the quality of the work, which indicates the worries and discussions of those days.

In this work, the space of a performance is approached as a place that creates meaning on the level of physical space, the space described in the text, and as spatial relationships between the performers and the spectators. The discussion uses the concepts of fictional and factual space, the performing situation and performance world. The performance–audience relationship is seen as composed of the stage-auditorium relationship in the space, the mode of address in the text, and the chosen performer–spectator relationship. Arnold Aronson's scale from environmental to frontal arrangements of space,[9] and Peter Eversman's model for analyzing theatrical space based on the structural dimension and the dimension of use,[10] are applied to example performances. The model is criticized on the basis of those experiences. The practical work is described as a process from text to space as well as from place to text. The use of space is treated in connection with questions related to director, actor, and writer points of view.[11]

The main premise of the work was that a live performance takes place as a space. My aim was to show that in creating a performance, the space can be an interesting starting point both in terms of spatial relationships and as a place creating meaning. In 1999, at a Nordic conference,[12] I chose to present the work in two parts, first, as a discussion with theatre researchers, then as a description of artistic development, in order to clarify my critique and contribution.

Acta scenica 3 – *Art, Narrative and Identity*

The third research publication at TeaK was an anthology of papers on cultural studies.[13] In my contribution I compared my research in *Some Conversations XI* (1995) and *Performance as Space* (1998). Both dealt with the space and place of a performance by producing artistic work and reflecting upon it in a report, though very differently. The text describes how I found a dichotomy between artistic and scholarly research artificial at first – to investigate and explore is possible in all areas – and how I later realized the usefulness of distinguishing artistic aims and research goals when the work contains both. Some remarks concerning artistic research were included, to distinguish it from demonstrations of excellence or from artistic development work:

> As far as I understand, research, also in a wider sense, requires at least two more things: First, you should position your work within some tradition or framework, in order to continue, add to, oppose, criticize or challenge it, even if you feel your work to be unique and unprecedented. This simply means that you should try to find out what has been written about your subject. However, in [Finnish] theatre originality and authenticity are often emphasized. Everyone is supposed to invent the wheel herself, preferably her own wheel, too. [...]
>
> Secondly, you should describe your experiences so they can be shared, to some extent, within the discourse you position yourself in. And you should generalize part of the experiences to concern others, at least some. [...] what is trivial and self-evident for you can be new knowledge for the reader.
>
> The third requirement is that you have courage to question and to experiment. This was my addition, my demand to myself. According to my experience, research and experimental performances go together. Unlike natural sciences or marketing, [where risk is minimised] [...] it is not accurate to speak of research in art if there is no risk, nothing unknown, no challenging question involved. What is the point of an experiment in art if you know the answer in advance? [...] Some uncertainty with regard to the result, risk taking, trial and error or doing differently is part of research in making art within ordinary production processes. Should that not be the case even more in a research oriented practice? In the words of Beckett "to be an artist is to fail like no other dare fail" [...] But, on this road, too, we easily end mythologizing the role of the artist.[14]

Comparing my research processes: In the first project a quasi-experimental research question – how does space influence a performance? – was reflected in the name of the project, "Space and place as an element of a performance," and was later specified into exploring "the place of action in the text and the place in the city as starting points for creating a performance world."

In the second project I chose an open problem instead of a question, "the performer's space and the spectator's space," which gradually transformed into "performance as space." Today the topic that interests me is landscape, and the question I work with is: how to perform landscape?

Acta scenica 19 – *Ways of Knowing in Dance and Art*

Since the first Acta scenica publications much has happened. One of the recent publications is an excellent example of the situation today. Leena Rouhiainen, the editor of *Ways of Knowing in Dance and Art*, introduces the research project "Challenging the Notion of Knowledge: Dance, Motion and Embodied Experience as Modes of Reasoning and Constructing Reality" carried out by ten dance artists, pedagogues and scholars from 2005 to 2007:

> By investigating various somatic strands of artistic practice in dance, the project aimed to illuminate the nature, meaning and possibilities of embodied knowledge. In doing so, it also searched for means to utilize this knowledge as a tool of investigation and creation for both scholarly and artistic ends. The overall aim of the project is to substantiate the paradigm shift towards a holistic notion of knowledge and to affirm the body as integral to the process of knowing.[15]

The dichotomy between artistic and scholarly/research dissertations has been abolished at TeaK, but many dissertations still have a scholarly emphasis. And a crucial choice still remains: Are the artistic parts evaluated as research results, or are they considered to be material, data for qualitative research or tools, a method of research, and treated only in the written report, that is, not considered as research outcomes? If the artistic parts are used as data, this resembles ethnographic research, where the researcher uses her or his own experiences as material. This is what practice-based research mostly has come to mean. Only scholarly oriented researchers working mainly with archival knowledge would probably find problems with that. The same applies when artistic practice is used as a method, which can be explicated, complemented by other methods, and the research outcomes discussed on the basis of the written report. New questions and interesting problems occur when artistic work is evaluated as research outcome, as part of the dissertation.

Artistic research can be understood as a form of research in itself, different from, but possible to juxtapose with, other forms like philosophical, historical, ethnographic research, etc. However, if we are not constantly focusing on the artworks (or artistic practice) and their role in the research project, then the specific knowledge embedded in artworks, artistic practices, and the artists themselves is easily bypassed, colonized, or assimilated into familiar forms of research. On the other hand, a dichotomy based on the idea of equivalence, which originally meant to strengthen the position of artistic work, can easily develop into an apartheid situation where artistic research has some privileges based on the traditional freedom of art, but no access to knowledge production or right to be considered as knowledge in and of itself.

Strict dichotomies between art and research, art and scholarship, art and science, or art and theory, all lead to absurdities and tend to devalue embodied and tacit knowledge. Contemporary art is often involved in various forms of knowledge production. And it is not only the knowledge of artists that is embodied and situated. Or so it seems from a Helsinki perspective.

Notes

1. I also work as an artist, *see* www.harakka.fi/arlander
2. *See* www.teak.fi/Research
3. Dissertations including examined performances are: Annette Arlander (1998), Riitta Pasanen-Willberg (2000), Betsy Fisher (2002), and Kirsi Monni (2004). Recently Helka-Maria Kinnunen (2008) has used her own performances as data for narrative analysis, but the performances were not examined as part of the dissertation.
4. Pentti Paavolainen and Anu Ala-Korpela, eds, *Knowledge Is a Matter of Doing: Proceedings of the Symposium Theatre and Dance Artist Doing Research in Practice*, Acta Scenica 1 (Helsinki: Theatre Academy: 1995).
5. Jerzy Grotowski, "Performer/Le Performer," in *Hän Ei Ollut Kokonainen – Tekstejä Vuosilta 1965–1969 (He Was Not Complete – Texts from 1965–1969)* (Helsinki: Painatuskeskus, 1993), 242.
6. *See* Mika Hannula, Juha Suoranta, and Tere Vadén, *Artistic Research – Theories, Methods and Practices* (Helsinki, Finland: Academy of Fine Arts; Gothenburg, Sweden: Academy of Fine Arts: University of Gothenburg/ArtMonitor, Gothenburg 2005); Henk Borgdorff, *The Debate on Research in the Arts* (Bergen, Norway: Kunsthögskolen i Bergen, 2006); Satu Kilijunen and Mika Hannula, eds, *Artistic Research* (Helsinki: Fine Art Academy, 2002).
7. Annette Arlander, "Some Conversations ... in Various Spaces," in *Knowledge Is a Matter of Doing: Proceedings of the Symposium Theatre and Dance Artists Doing Research in Practice*, ed. Pentti Paavolainen and Anu Ala-Korpela, Acta Scenica 1 (Helsinki: Theatre Academy, 1995), 118–19.
8. Ibid., "Joitakin Keskusteluja Xi, Raportti Tee -Projektin Esityssarjasta, Joitakin Keskusteluja I-X (Some Conversations Xi – a Report on the Series of Performances Produced by Tee-Project, Some Conversations I-X)" (Helsinki: Theatre Academy, 1995).
9. Arnold Aronson, *The History and Theory of Environmental Scenography* (Ann Arbor: UMI Research Press, 1981).
10. Peter Eversman, "The Experience of Theatrical Space," in *Performance Theory, Reception and Audience Research*, ed. Henri Schoenmakers (Amsterdam: Tijdschrift voor Theaterwetenschap/ICRAR, 1992).
11. Annette Arlander, *Esitys Tilana (Performance as Space)*, Acta Scenica 2 (Helsinki: Theatre Academy, 1998), 7.
12. Annette Arlander, "From Space to Place to Landscape" in *Further and Continuing Education of Performing Artists in the Nordic Countries – a Nordic task*. Proceedings from the conference in Oslo, 12–14 November 1999 (Nordic Theatre and Dance

– Nordic Centre for the Performing Arts, Nordic Council of Ministers, TemaNord, 2000), 621.

13. Pia Houni and Pentti Paavolainen, eds, *Taide, Kertomus Ja Identiteetti (Art, Narrative and Identity)* Acta Scenica 3 (Helsinki: Theatre Academy, 1999).
14. Annette Arlander, "Tila Ja Aika Eli Miten Megalomaanisesta Idealistista Tuli Suhteellisuudentajuinen Pragmaatikko (Space and Time – or How a Megalomaniac Idealist Turned into a Pragmatist with a Sense of Proportion)," in *Taide, Kertomus Ja Identiteetti (Art, Narrative and Identity)*, ed. Pia Houni and Pentti Paavolainen, Acta Scenica 3 (Helsinki: Theatre Academy, 1999), 51–3.
15. Leena Rouhiainen, ed. *Ways of Knowing in Dance and Art*, Acta Scenica 19 (Helsinki: Theatre Academy, 2007), 5.

10

An Actor Prepares: Performance as Research (PAR) in the Theatre

Ian Watson

The gold standard of research marches to the tune of science. Its harmonies are hard evidence, quantifiable data, objective facts, and findings that can be reproduced. Art, rooted in craft, the aesthetic, personal experience, and the embrace of contradiction, seems far removed from this technology of discovery.

Thankfully, the world is not short of individuals who dare to marry a research agenda with creative enterprise: men and women who combine the artistic complex of inspiration, talent, training, skilled craftsmanship, experience, and personal expression with a deep investigation of their artform that bears the hallmarks of academic rigor. This is hardly a world of white coats, test tubes, or even libraries. It is much closer to the fieldwork of anthropologists or sociologists, except that in this case the performers' actions are the subject of their own investigations. Theirs is a form of what the sociologist Kurt Lewin termed action research, that is, research conducted upon current activity by those involved in the activity to better understand it and to develop strategies to improve current practices.[1] Action research in the arts explains the investigations conducted by Jackson Pollock as he developed his action painting techniques or Stravinsky's various experiments with musical motifs or his studies of folk materials as compositional sources. It also sheds light on the work of theatre artists such as Stanislavsky and Grotowski, whose painstaking analysis of the actor's craft shifted the valence of theatre from the text to performance during the last century.

The theatre paradigm

In the essay "Methodical Exploration," Grotowski compares the work he did with his actors in his theatre years (commonly referred to as his "art as presentation" period) to work at the Bohr Institute at the University of Copenhagen in Denmark.[2] The institute, named for the famous Danish theoretical physicist Niels Bohr, is a major scientific institute devoted to the study of atomic and quantum physics.[3] While acknowledging the differences between science and the theatre, he argues for an exploration of the actors' craft that parallels scientific research and a physical home that mirrors Bohr's institute.[4] In claiming comparisons with the benchmark of serious research, Grotowski gives voice to an undercurrent in the theatre that both preceded and followed him: that of performance as research.

Stanislavsky is the most widely known of those who embraced performance as research. Unlike those who came after him, his work was highly successful in the mainstream. He formulated a basic process that has proven flexible enough to survive numerous adaptations since its inception, and his research was either concerned with or readily adaptable to a system of acting that could be applied to a range of production genres with a great deal of success. The various modifications of his system are relatively easy to learn; and, most importantly, they can be taught by a trained instructor generations removed from Stanislavsky himself. Then, of course, there are the books, though hardly documents that traditionalists would accept as research, they are the product of many years of investigating the acting process. They suggest performance strategies and provide advice to actors; they articulate Stanislavsky's thinking about the theatre as well as his understanding of theatrical presentation, and confirm his findings in the conventional language of research, the written word.

Stanislavsky's books are only part of a larger canvas. The historical valence of his investigations and influence on those that succeeded him is measured by the legacy of his larger *oeuvre*. Unlike scientific research, which is grounded in the discovery of universals, performance research is rooted in the personal, in the findings of individual actors and/or researchers in particular circumstances which are, in turn, applied in future situations. The findings are not about the revelation of natural laws; they are confirmations of particular performance research methodologies and practices that can be applied on stage as well as passed along to others in the studio. Their legacy is confirmation that the findings of performance research are not limited to the idiosyncratic and that they continue to impact upon current performance practice.

The combination of the personal and the ephemeral in the application of performance research findings lies somewhere between good advice and truth. Even though Grotowski's early pupil and champion Eugenio Barba – who spent many years studying performance across cultures with his actors at the Odin Teatret and at the International School of Theatre Anthropology (ISTA) – talks of the laws of performance, these are hardly laws in the same way as those that govern the natural world are laws. They are laws of pragmatism, learned in acting classes, in the workshop studio and on stage, that provide a foundation for each performer who applies them. Stanislavsky and others, like Lee Strasberg and Stella Adler, who developed actor-training systems based on their interpretation of his discoveries, provide a grammar of performance designed to transform the playwright's words into performance. But this grammar is a guide rather than a set of rules that must be followed precisely. The grammar is shaped by each actor's initial training, professional experience, ongoing study with master teachers, personal insights, and the particular project he is working on. This personal component is even more pronounced in the work of artists like Grotowski and Barba, whose research paradigms deny acting systems, preferring instead to develop methodologies of performance research focused on subjective investigation; these personal discoveries, that vary from one actor to another and become gradually embedded in the actor's body through years of application, are the premises from which performances are created and sustained.

Stanislavsky's continued influence on actor training and performance, Grotowski's importance to those who approach acting in a certain way, and Barba's impact on group theatre demonstrate that their discoveries resonated beyond the confines of Moscow, Wroclaw, and Holstebro. Their legacies incorporate both a contemporary and historical dynamic. The former encompasses the relationship between research findings and application as well as between knowledge formulation and teaching; the historical component is the inventory of completed research that informs performance and is continually being enriched by new hypotheses and results – what Grotowski termed the collective memory of theatre.

Training and research

The lines between training and research in theatrical performance are often unclear; what is training to some, is research to others, and vice versa. It would help if one could say with confidence that professionals research while students train; but things are hardly so simple. Professional actors often take classes with the likes of Anne Bogart, Roberta Carreri, or Gennadi Bogdanov. Understandably, students are less likely to engage in research than their professional counterparts are, but they do in places such as the Experimental Theatre Wing at New York University, the Odin Teatret, and Poland's Gardzienice.

The emphasis in training is on the preparation for performance. If one studies at the Odin Teatret, Gardzienice, the Actors Studio, in a university-based conservatory, or with one of the many private teachers in New York or Los Angeles, the attention is on learning and assimilating the technical skills that can then be applied directly to stage and/or screen performance. Performance research is less directly focused on preparing a play or film role for presentation to the public. It is about exploring possibilities, testing limits, and laying foundations that inform but do not constitute public performance.

Embodied research

Despite the differences among Barba, Grotowski, and Stanislavsky, their research models have surprising similarities. For each of them, the research begins from a challenge (such as how to establish the relationship between acting principles and the personal for Barba; Grotowski's attempts to mine the psyche and transpose the process into a *mise-en-scène*; and Stanislavsky's need to identify the strategies that allow an actor to tap the spontaneous at will, while remaining engaged in the moment to moment unfolding of the play's circumstances). These challenges generate a series of explorations (experiments in scientific terms) in which fruitful lines of inquiry are pursued while failed lines are rejected, and findings are made, tested, and, if proven worthy, are applied. Often, these findings become the basis of further research.

The general layout of this model would be recognizable to most science students, but there are, of course, major differences between science and the creative process being investigated by those concerned with acting. The primary concern for the latter is enriching the act of performance. The researcher is the artist herself, she tests possible avenues of inquiry and analyzes tentative conclusions; documentation is primarily corporal, that is, retained in her body; and the application of findings is equally her domain.[5]

This embodied research model denies both the tenet of objectivity and the reproduction of findings by other parties that is so central to scientific discovery. It also serves to highlight the tacit component of documentation. Corporal findings based upon active research, the repetition of physical and/or vocal patterns, and psycho-emotional constructs frequently become part of the actor's body memory. Applications of findings in these instances often engage this body memory without regard to the conscious mind. An experienced driver, for example, does not go over every aspect of driving in his mind while in charge of a vehicle. The body remembers the tasks required, freeing him to concentrate on other things. Walking is much the same for most of us, as is a well-learned score for an actor or dancer.

Embodied documentation supersedes the literary in this auto-research process. Even Stanislavsky's research findings, which he documented himself and have been formulated into a system of acting, largely by others, is essentially embodied. Performers do not learn how to act on stage from books. Stanislavsky-trained actors study in studio/workshop settings with teachers whose techniques are rooted in one or other formulation of his findings, which they in turn have learned from someone else; the master's writings, if read at all, serve as reinforcement to the process of active learning that is transmitted orally.

The genealogy of oral transmission lies at the heart of Stanislavsky's rejection of his approach to acting being regarded as a system, a rejection that has broader implications for understanding the training/research dichotomy in mainstream acting. Stanislavsky's original formulations were based on a combination of his own personal research and the research undertaken by colleagues, such as Leopold Sulerzhitsky in the Moscow Arts Theatre's First Studio. Stanislavsky continued to work on and adjust his ideas on acting throughout most of his life. Nevertheless, the functional success and emotional logic of his findings and those of others who have developed their own approaches to actor training based on interpretations of his formulations, such as Strasberg, Adler, and Meisner, have popularized the notion of acting systems. This notion lends itself to closure because it suggests a combination of techniques and principles that together form a unified whole that privileges application over further interpretation. Stanislavsky seemed more inclined to an open process. One in which personal research plays an important role. As far as he was concerned, actors should repeatedly test what they have learned. They apply what works for them and reject what does not. Professional experience, on stage, on the set, and in the workshop/studio setting, is the laboratory for the conventional actor. To be fair, the differences between Stanislavsky and his American interpreters may have more to do with emphasis than incompatibility. For Strasberg et al., the priority is on the system; for Stanislavsky it appears to be on how the individual actor applies and continually adjusts the system. Stanislavsky's emphasis has more to do with personal research than mastering a technique of truths, which, as others have pointed out, is what happens in practice, even on an American stage dominated by the propriety of specific techniques.

Legacy and performance research

The collective memory of theatre is a weave of embodied, oral, and literary texts, of deep reflection and experiments, challenges and failures, heavy lifting and successes. Most of all, it is the work of individuals committed to the collaborative nature of the theatrical enterprise. Playwrights loom large in this history because they most often provide the raw material of presentation, a few scholars also have their place, but those that have infused the transition from the written to the spoken word in empty spaces are of greatest importance. The investigations of Stanislavsky, Grotowski, and Barba have not only filled their own stages with exciting alternatives to the deadly theatre of predictable habit, but have also influenced their contemporaries and the generations that followed them. These influences are the legacy that confirms the importance of the performance as research paradigm.

Any legacy of performance research demands a process of transmission and the survival over time of the transmitted knowledge. The latter is history, the former the function by which it is written. The legacy is made all the more difficult to parse because the personal discoveries that drive it are embedded in the bodies that conducted the research rather than in an objective, systematized compendium of performance techniques that can be easily taught and/or accessed through reading books or watching films about them.

The transmission of performance knowledge is complex. This is because, unlike more conventional research, which privileges material documentation in one form or another, performance knowledge is ephemeral and idiosyncratic. Some results of performance research are documented in written form, but as influential as books such as *An Actor Prepares*, *Towards a Poor Theatre*, or *Dictionary of Theatre Anthropology* have been and continue to be, it is the study with others that has the highest transmission quotient for performers. This direct form of transmission, in which performance knowledge based on a combination of learned technique and personal research is passed from teacher to pupil, from skilled performer to neophyte, is predicated upon the teacher/craftsperson performer having absorbed a corpus of performance knowledge into her body. This knowledge is a text of sorts, a text that is not only a repository of information and experience, but is also an active site of its application; in other words, she knows both the "what" and "how" of her subject.

Yet performance knowledge is amorphous. It can be talked about, even committed to paper, but it only exists in the doing. It encompasses one's initial and ongoing training through workshops, scene study classes, study with performers from another culture, life experience, and so on, that an actor is involved in during her career, as well as what she learns about applying her training during performance. It is a fusion of training, interpretation, and experience in which the various genesis strands are difficult, if not impossible, to tease apart. This is a history of continual accretion grounded in a core of basic training and personal research passed from master to pupil. The accretion is in a constant state of flux while the teacher and student remain engaged artists, however, because they continue to be exposed to new experiences, ideas, and techniques that further shape their ongoing research and the performance knowledge it garners. A young actor studying at the Actors Studio, the Atlantic Theater School or at the Adler Studio today receives a training rooted in variations of the research Stanislavsky conducted in the early part of the twentieth century, filtered through the experiences and research of the various teachers and pupils it passed through on the way. Applying the technique in performance, further study with master teachers, and life experience build upon and reshape this initial generic training to formulate a personalized work process. Should the actor become a teacher, it is this personal shaping of the technique that forms the basis of what is shared with students.

This lack of precision is exacerbated by the role of embodied documentation in the process. Much of what has been discovered and stored in a performer's body is tacit rather than explicit knowledge. It resides in the body memory with little need for articulation and is passed from performer to pupil by strategies such as combining generic exercises designed to prepare the actor to learn and apply techniques in studio settings (Stanislavsky-based schools such as the Adler and Actors Studio), imitation (as in much of Asia), or by the master performer providing a vague premise as a starting point in an exercise to a pupil, then guiding them through it towards a goal situated somewhere between what they know intuitively and what has to be discovered anew by the pupil (the Odin workshop method).

A collective memory

The transmission of performance knowledge and the legacy of those who have formulated it is a complex and ongoing process. This process of accretion and its history in the bodies of performers is the collective memory that Grotowski claimed for the theatre. A living history that confirms the historical importance of performance research among at least the three artists I have focused on here. Its importance is not, however, limited to these central figures. Theirs is a particular history, born of an era they helped shape; but they are part of a historical process that began with those who prepared the way for Thespis as well as Bharata and which continues into tomorrow as long as theatre breathes life into empty spaces anywhere.

Notes

1. Kurt Lewin, "Action Research and Minority Problems," *Journal of Social Issues* 2, no. 4 (1946).
2. Grotowski's "art as presentation" period marks the first of several phases in his career, see Jennifer Kumiega, *The Theatre of Grotowski* (London and New York: Methuen, 1985).
3. Jerzy Grotowski, *Towards a Poor Theatre* (New York: Simon & Schuster, 1968), 127–32.
4. Grotowski's original theatre company – the Theatre of the Thirteen Rows – eventually incorporated the term laboratory into its name and in March 1962 was renamed the Laboratory Theatre of Thirteen Rows. When recognized by the Polish state in 1966, the company's name was changed to Theatre Laboratory "Thirteen Rows" – Institute of Research into Acting Method.
5. Documentation can also be textual; several of the Odin actors maintain research journals on what they are doing and what they discover. In turn, the journals become a source for continued research.

11
Making a Dance/Researching through Movement

Susan Leigh Foster

Since its introduction into university curricula, the subject of dance has been conceptualized as providing a unique opportunity to experience and create knowledge unavailable in other forms. Central to this argument has been the practice of creating movement and from that, new dance. From the first courses introduced into physical education programs around the United States at the beginning of the twentieth century, students were taught ways of moving as well as ways of creating and sequencing movement in order to produce a dance. Both the new ways of moving and the dances that developed out of them, though never labeled as such, were always envisioned as research. For example, the University of California has long recognized "Creative Activity" along with "Scholarly Publications" as viable formats for the production of new knowledge.

In order to elucidate what this research process might consist of, this chapter examines three distinctive approaches to dance making developed by three highly influential composition teachers: composer Louis Horst, who worked at Bennington College in the 1930s; choreographer Daniel Nagrin, who frequently taught at the American Dance Festival in the 1950s and 1960s; and accompanist and movement analyst Robert Ellis Dunn, who hosted classes at the Cunningham Studio in the 1960s. Versions of their approaches have permeated textbooks and syllabi for composition courses in universities throughout the latter half of the twentieth century. Conducting a comparative analysis of their approaches, I want to emphasize the distinctiveness of their methods so as to highlight the diversity of possibilities for undertaking practice as research in dance.

Research as moving

Before focusing on the teaching of dance composition, I turn to the pioneering efforts of educator Margaret H'Doubler, who was hired in the Women's Physical Education Department at the University of Wisconsin to develop a course of study in Dance in 1917.[1] After several years of diligent expansion to the curriculum, H'Doubler saw her program awarded the first BA in Dance within Physical Education in the United States in 1926. As rationale for including dance in the university curriculum, H'Doubler argued that dance is the translation of emotional experience into external form. In order to accomplish this transference, the body's responsiveness as a physical mechanism must be mastered.

H'Doubler advocated a kinesiological understanding of the body's movement capabilities as a way to understand its propensity to move. Often working blindfolded, her students were asked to explore the range of motion at each joint, based on their study of the human skeleton. Abhorring any pedagogical approach based on imitation of movement routines, H'Doubler arranged classes so that students improvised most of the movement rather than copying movement performed by the teacher. Eventually, this training would enable students to produce art, defined by H'Doubler as the free translation of internal emotional experience into external bodily form.[2] Thus each student "researched" her own impulses to move, and only her anatomical proclivities to realize such impulses.

To her scientific exploration of bodily capacity, H'Doubler added sessions in which students collaborated, under the teacher's guidance, on the making of a dance. For their first experiments in learning composition, she argued, students could work on devising movement sequences for a select piece of music. She explains:

> When the phrasing is understood, have the class skip to the right for one phrase [...] Then ask the class what to do next. Some will suggest going on in the same direction for another phrase; others will recommend going back to the left. Try both. The class will discover that skipping back for a phrase gives balance. Now ask the class if they have a satisfying sense of completion, or if they feel the need of repeating what has been done. Of course, some will want to repeat. So this should be done.[3]

Not unlike a rudimentary laboratory experiment in chemistry, H'Doubler guides her students through the process of exploring different effects that result from the addition of different materials. Her own aesthetic preferences, masked beneath the investigatory rubric of trying different options, clearly led students in a certain aesthetic direction, cultivating their ability to craft phrasing, floor path, and ensemble shapes, and from these discoveries, students could extend themselves further into composition. However, the underlying intimacy established between idea and action continued to inform their explorations, and the potential for this inquiry to yield new insights would undergird dance pedagogy for decades. Even as the study of dancing and dance making acquired greater regimentation, both in terms of the specification of techniques for training the body and programmatic approaches to the study of composition, this conception of physicality as resource-full continued to guide the study of dancing.

Research as composing

In response to the growing popularity of modern dance on the concert stage, as well as the pedagogical efforts of educators such as H'Doubler, Bennington College established a summer program of study in 1934, one which attracted an impressive number of students who went on to become university educators of dance. The curriculum distinguished among technique, composition, and choreography courses. Where H'Doubler asked students to improvise much of the movement generated in class, by the 1930s, one trained to become a dancer using a more prescribed regimen of exercises, often devised by the choreographer, that, on one hand, exemplified the choreographer's aesthetic vision, and on the other, embodied "universal" principles of motion. Where H'Doubler based her course plans on the body's kinesiological capacity to move in any and all ways afforded by its structural organization, Bennington technique classes, taught for example by Martha Graham, Doris Humphrey, and Hanya Holm, proposed distinctive sets of principles as the underlying foundation for dance movement. Research now took place more in composition and choreography courses than in technique courses.

In composition courses students learned structuring principles that imparted an ability to analyze movement in terms of space, time, and weight. Envisioning space as a void into which the body projected various shapes and energies, and time as a measure of the quickness or slowness of motion, they activated a momentum-filled relationship between central and peripheral body, creating short studies that demonstrated their understanding of the possibilities for shaping the body as a three-dimensional object in space, and for sequencing those shapes according to various musical structures. Students did not, however, create "dances," envisioned as the development of a vision or argument in movement. Only in the choreography workshop could these students witness and assist in the birth of a dance, one whose thematics depended upon the inspiration of an already established artist. Both studies and dances, however, were envisioned as outcomes of a hyper-personalized process wherein the individual became origin of the movement, host to the creative process, and craftsperson of the dance's development.[4]

It was under the rubric of composition courses that Louis Horst, composer, accompanist, and adviser to Graham, introduced his Pre-Classic Forms and Modern Forms courses, which were taught regularly through the 1960s at the successor to Bennington, the American College Dance Festival. In Modern Forms, Horst first proposed studies based on simple structures typically used in music composition such as theme, contrasting theme, return to theme, ABA, or the rondo form ABACADA. These kinds of structures enabled students to explore how movement could be developed from an original phrase into its repetition, inversion, amplification, or contraction. Using that basic structure, he asked students to strengthen their knowledge of, and skills at, developing three aspects of movement – spatial design, rhythm, and texture – by creating studies that explore strange, asymmetrical positionings of the body or where each body part activates a different plane; asymmetrical time structures such as 5/4 or 7/8; and dissonant and contrasting textures.

Having acquired facility at these conceptualizations of movement and its development, students would embark on a series of studies based on distinct styles. According to Horst, each style informs the modernist aesthetic in a crucial way and is tied to a kind of individual and social temperament, so that mastery of that style will yield tools for representing a wide variety of psychological states. These states are, in turn, aligned with various regional, ethnic, or social configurations, so that once students gain fluency with the styles, they can make studies that fuse the personal with the social.

For each style, Horst provides a succinct summary of movement qualities and characteristics.[5] In "earth primitive," for example:

> The dancer is alertly sensitive to the feel of the earth under his feet. It is the genesis and grave of all living things – the source and the finale. The movements are in the lower areas and oriented to the floor. They can be clumsy and animalistic. They can be brutal and threatening. They can project the lyricism of wonder, or the tenderness of the giver of life. They may have a drum-like percussiveness. But always they are simple and meagerly articulated; lean and taut.[6]

Here the notion of the primitive is mined for the spatial and temporal characteristics associated with stereotypic images of it. Horst draws from these characteristics psychological attitudes such as brutality or tenderness, and finds connections between movement and psyche in word choices such as "lean" and "taut."

In addition to the more historical and psychological forms, Horst added styles associated with a particular people. For example, Jazz style "retains attributes brought from Africa: jerky, percussive movements and accents. It displays qualities of syncopation which grew out of a disintegrated people: a melancholy and lassitude, resultant of slavery."[7] "Americana" consists of "big, free and extended movement. [. . .] It is swinging and out-going – a wide stance in the legs and an open carriage of the chest."[8] As these latter examples make especially clear, Horst conflated individual and cultural stereotypes, assuming that personality or temperament could stand in for or symbolize key cultural values and ways of being. The well-made study would transcend the stereotypes through its innovative interpretation of both the individual and the cultural and also through its mastery of movement's thematic development.

Students trying to fulfill one of these assignments probably worked partly in front of a mirror, evaluating their production of asymmetry, parallel lines, or a sufficiently wide stance. They might also try out a number of asymmetrical positions or possible stances in order to determine which might be more innovative, persuasive, or cohesive with the movement phrase as a whole. These periods of investigation and assessment, mostly referred to as "being in the studio," did not attract scrutiny or discussion. Rather, the studies themselves, as the outcome of these periods of investigation, served as objects of analysis. As such, they were evaluated in terms of the acquisition of compositional skills and originality.

Researching the psyche's movement

In contrast to Horst's formal strategies for representing psychological and cultural states of being, Daniel Nagrin's approach to composition provides tools for drawing the initial impulse for a dance out of the psyche and expanding it into character and narrative.[9] Whereas Horst believed that aspiring artists should first master a range of compositional techniques before creating a dance, Nagrin focused the student from the beginning on a rigorous introspective inquiry. In his book *Dance and the Specific Image,* he explains that he begins each new composition by identifying a personal conundrum: "I start out with an intense focus on some part of existence that engages me, baffles me and demands attention from me."[10]

Using improvisation, he begins to explore a specific motivation or situation – what he calls an image. The image, or some part of it, informs each action. Much of the process of dance making entails a rigorous scrutiny of the extent to which each moment in the dance, each motion of the body, fulfills this image. Where Horst tended to uphold the distinction between composition and choreography, Nagrin assumed that students could begin right away to make dances through the process of continually honing the relationship between image and action.

In classes, Nagrin engaged students in many different kinds of improvised scenarios that would inspire a discovery of a specific conundrum and/or assist them in developing the core idea. He might ask students to consider one of the following:

> Someone/something is leaving. You are the one leaving or the one being left.
> Someone/something is protecting and keeping someone/something safe and warm. Are you the protector or the protected? [. . .]
> There is some something/someone of which you are in awe. Become that someone/something.
> There is something/someone which fills you with fear and loathing. Become that something/someone.[11]

Nagrin asked students to explore these kinds of themes in class and also in their individual rehearsals in order to find an image that could be developed into a dance.

Once students develop an image, Nagrin asks them to consider six basic questions that should guide the work towards completion: (1) who? or what?; (2) is doing what?; (3) to whom? or to what?; (4) where/when?; (5) to what end?; and (6) engaging what obstacle?[12] Where Horst envisioned structures such as ABA or spatial contrast as the necessary tools to develop a dance, Nagrin relies instead on the analysis of psychologically driven narrative. To the extent that his questions are answered fully and honestly, the dance will offer an acute and vivid rendition of the predicament. Whereas Horst teaches mastery over formulae designed to evoke psychological and cultural dispositions, Nagrin focuses on the relationship of individual motivation to action. Movement is seen less as a medium to develop and not at all as a substance that is significant in and of itself. Rather, movement is utilized to create character and story.

Researching the process of arranging movement

Where Horst and Nagrin focused on the psychological and interiorized selfhood of the choreographer, Robert Ellis Dunn concentrated precisely on movement, distinct from any psychological or social significance, as material to be developed by making specific decisions concerning parts of the body, timing of actions, and locations of events. A student of John Cage and accompanist for Merce Cunningham's technique classes, he began teaching composition in 1960 at the Cunningham studio to a group of artists who became known as the Judson Dance Theatre group. Although radically different from Horst and Nagrin in eschewing any connections between movement and psyche, Dunn nonetheless tapped the foundations of modern dance as research established by H'Doubler. Like her, he focused on possibilities for the body to articulate itself.

One of his first assignments, according to Simone Forti, "was to make a dance by combining sets of choices for body parts, durations, parts of the rooms, and left or right directions in space."[13] Following Cage, he embraced any and all sources of movement as viable vocabulary for dance. This approach to dance composition, relying on the arbitrary choices of a depersonalized procedure for making choreographic choices, profoundly alters the alignment between motivation and action, and radically redefines authorship. It encourages its practitioners to look on composition as a process of continual experimentation, where each set of choices produces certain results.

This focus on process, likewise, permeated discussions of the students' projects. As Trisha Brown recalls:

> After presenting a dance, each choreographer was asked, "How did you make that dance?" The students were inventing forms rather than using

> the traditional theme and development or narrative, and the discussion that followed applied non-evaluative criticism to the movement itself and the choreographic structure as well as investigating the disparity between the two simultaneous experiences, what the artist was making and what the audience saw. This procedure illuminated the interworkings of the dances and minimized value judgments of the choreographer [. . .].[14]

Rather than privilege the formal design, contrasting elements, and rhythmic clarity of a dance, as Horst would, or the characterological development, as Nagrin would, Dunn placed great value on how the compositional process, whatever it was, revealed itself in a certain structuring of the product. His interest in movement as physical articulation opened composition up to a much wider variety of sources and approaches while at the same time disavowing motivation, character, and narrative. It also emphasized the potential for choreography to function as an ongoing site of inquiry and investigation.

Each of these approaches to teaching composition establishes what a dance is and how it should be made. And each offers a distinctive conceptualization of dance making as a form of research that includes the potential for reflexive analysis of the research process itself. Horst's approach invites the individual to historicize his or her own aesthetic situation, to identify styles and aesthetics that inform one's own historical moment, and to probe the relationship between the self and the social. Nagrin's approach demands an unrelenting scrutiny of one's motivation and a detailed evaluation of the accuracy of its representation. Dunn's approach encourages experimentation as a willingness to try anything and everything – to assess what differences are entailed in each effort and to note the results of a given decision, and then to keep track of those correlations as one tries again and again. All three, when examined in detail, make clear how research in dance can occur within a variety of formats and with diverse epistemological assumptions. At the same time, all three affirm H'Doubler's assertion that the body's movement can and does produce knowledge.

Notes

1. For a fulsome analysis of H'Doubler's life and work, *see* Janice Ross, *Moving Lessons: Margaret H'doubler and the Beginning of Dance in American Education* (Madison: University of Wisconsin Press, 2000).
2. Margaret H'Doubler, *The Dance* (New York: Harcourt, Brace, 1925), 11.
3. Ibid., 172.
4. Both the invention and sequencing of movement were seen as each student's creative investigation, unlike scientific inquiry, which is typically conducted by a team.
5. In addition to "Earth Primitive," Horst identified "Air Primitive," "Archaic," "Medieval," "Introspection," "Jazz," Americana," and "Impressionism" as styles informing the Modern.

6. Louis Horst and Carroll Russell, *Modern Dance Forms in Relation to the Other Modern Arts* (Brooklyn: Dance Horizons, 1977), 61.
7. Ibid., 111.
8. Ibid., 126.
9. Daniel Nagrin's approach to composition developed first out of his long-standing association with choreographer Helen Tamiris, one of the most renowned modernists, who began choreographing in the 1940s. Nagrin began teaching his approach to composition in the late 1960s, first at the State University of New York at Brockport, then regularly at the American Dance Festival summer program, and in guest residencies around the USA.
10. Daniel Nagrin, *Dance and the Specific Image: Improvisation* (Pittsburgh: University of Pittsburgh Press, 1994), 31.
11. Ibid., 49.
12. Ibid., 30.
13. Quoted in Sally Banes, *Democracy's Body: Judson Dance Theater, 1962–1964* (Ann Arbor: UMI Research Press, *c.*1983), 11.
14. Quoted in ibid., 20–1.

12
Performance as Research (PAR) in North American Ethnomusicology

Sandra Jean Graham

Ethnomusicologists who participate in musical performance as a research methodology gain unique perspectives on aesthetics, cognition, social relations, identity, and social structure. This chapter surveys the benefits and challenges of performance as a mode of research in North American ethnomusicology in the twentieth and twenty-first centuries.[1]

The roots of ethnomusicology extend back to the nineteenth century, when its practitioners relied primarily on the collection and quantification of performances that were frozen in recordings. Research goals gradually focused on performance in its cultural context during the twentieth century under the influence of anthropology (Richard Bauman, Milton Singer, Victor Turner), folklore (Roger Abrahams), linguistics (Noam Chomsky, Ferdinand de Saussure), and performance studies (Richard Schechner, Barbara Kirschenblatt-Gimblett).[2] By the 1950s, most ethnomusicologists had adopted participant-observation as their primary methodology and understood musical performance as *cultural* performance, although, as Carol Silverman notes, they often favored observation over participation.[3] Anthropologist Alan Merriam succinctly encapsulated the notion of musical performance as cultural performance in his famous definition of ethnomusicology as the study of music *in* culture, and later as the study of music *as* culture.[4]

Mantle Hood thrust performance into the spotlight when he encouraged university students to become proficient in another culture's musical tradition in order to expand their basic musicianship,[5] but this was of limited relevance to field research. In the work of British ethnomusicologist John Blacking and the American ethnomusicologists Marcia Herndon and Norma McLeod, however, researchers found ample motivation for using performance to understand not just the rules of musical style, but musical cognition.

According to Blacking, the first and most useful step toward discovering the principles that generate music and elucidate its relationship to other social activities and cultural forms is to "start singing and playing the music, and to invite criticism of one's performance" – something that his South African informants supplied in abundance.[6] In fact, Blacking decided that "the best way forward was to enter Venda society like a Venda child," learning children's music and being corrected by children.[7] For Blacking, the goal of performance was "to elucidate and understand processes and products," not to become a "virtuoso."[8] His interest in cognitive anthropology (which he viewed as a synonym for ethnomusicology) bears the influence of Chomsky's theory of generative grammar and Saussure's *langue* and *parole*.[9]

Herndon's perception of a musical occasion as a "nonverbalized expression of basic cognitive values and forms"[10] invokes Blacking's interest in cognition, Merriam's concept of music as culture, and Singer's concept of cultural performance.[11] The work of Herndon and McLeod is a frequent point of reference for those who use performance as research, which gained currency as a methodology in the 1980s.[12] The remainder of this essay addresses the benefits and challenges of performance as research in several key studies. In all of the following examples, the ethnomusicologist's participation in performance is only one in a constellation of research methods that might include observation, visual/audio/written documentation, interviews, archival work, laboratory analysis, transcription, and discussing conclusions with collaborators.[13]

Benefits of performance as research

Illuminating the invisible

In *Why Suyá Sing* (1987), Anthony Seeger is interested not in *what* Suyá music is, but *why* it is. Seeger and his wife spent 24 months living among the Suyá Indians in central Brazil, participating as widely as possible in daily life, from hunting animals to food preparation to social and ceremonial activities. The centerpiece of Seeger's ethnography is his description of the Mouse Ceremony, a rite of passage in which young boys are initiated into the male activities of the village plaza. It extends over two weeks and culminates in a 15-hour nighttime performance of singing and dancing in which men become mice-men (part animal and part human) and finally return to human form.

Participating in the ceremony illuminated aspects that would have remained invisible had Seeger relied only on observation. Frustrated by the Suyás' perfunctory descriptions of actions that were obscured by total darkness, Seeger discovers what really happens during one stage of the ceremony only when he participates in performance.[14] At another point in the ceremony, Seeger leaves the house in which the men have been singing by joining what he deems is a free-for-all rush for the door. He discovers that there is a designated order for departure when his partner pulls him back and "explains that our name set should never leave the house first, but instead should be among the last to leave."[15]

On a deeper level, Seeger arrives at an understanding of the invisible process of transformation through his participation in performance. After 15 hours of singing and shouting, leaping and dancing on the last night of the Mouse Ceremony, Seeger's group of performers is charged with the ritual obligation of continuing on long after the rest have stopped. "Our steps drag, our voices crack. The oldest name holder insists that we do not stop."[16] Finally the ritual ends – they die and are stripped of their capes. They are doused with cold water, remove all decoration, bathe in the river, and return to the village. The ritual death and the return to "real" time and place are much more powerful for Seeger having personally experienced them, and persuasively reinforce his argument that "space, time, the body, and social identity all defined, and were defined by, vocal art."[17] Music and dance did not merely reflect or reiterate other realities, but created them.

Understanding choice and creativity

Why do musicians make the choices they do? Seeger notes that the Suyá liked to experiment: "creativity was part of the fun of social life"[18] and it was applied to song as well as to speech, decoration, and mode of participation in ceremonies. Through participation, Seeger learned how mistakes, and their corrections, become part of the performance; consequently, performance is "real behavior" rather than an abstract ideal to which one aspires.[19]

In his ethnography of Bulgarian folk music as told through the biographies of Kostadin and Todora Varimezov, Timothy Rice is similarly interested in creativity, noting that "people not only play music, they play with music."[20] Rice used his experience playing the *gaida* (folk bagpipe) to understand individual variation within tradition, which meant ascertaining the cognitive processes that players used to generate their sound.[21] In instrumental traditions it is ornamentation that signals mastery, but older musicians and villagers especially lacked the vocabulary to describe the intricacies of musical experience. In his lessons, Rice "came to understand some of the meanings that Kostadin attributed to the ornamental notes," which "would have been inaccessible to an analyst who merely transcribed the notation."[22] Once Rice's mind and fingers learned how to execute the ornamentations technically, his "heart" had to give the music life through subtle variations, making the song his own.[23] Performance is an especially effective way of researching how play occurs in music and how audiences evaluate it.

Michelle Kisliuk also explores choice and creativity using a performance paradigm. Kisliuk came to ethnomusicology by way of New York University's Department of Performance Studies, and her research among the BaAka (forest people, or pygmies, of Central Africa) employs performance as research, as well as self-reflection on this strategy, to a greater degree than most. The first time she danced Mabo, she wrote, "My senses tingled; I was finally inside the singing and dancing circle [...]. This was different from listening or singing on the sidelines because, while moving with the circle, I became an active part of the aural kaleidoscope. I was part of the changing design inside the scope, instead of looking at it and projecting in."[24] As she moved around the circle, "the voices of different people stood out at moments," affecting her own singing and choices of variations.[25] On another occasion, she sang as she walked through the forest (as the BaAka commonly did), and "found that the close interaction of BaAka song with the surrounding forest weaves singing and listening into a simultaneous process"[26] – a point that was self-evident to the BaAka and therefore not worthy of comment by them.

Identity politics

"The politics of ethnicity in taiko are [. . .] bottomless, yet I remain sure that I learn something about Asian America when I play," writes Deborah Wong,[27] who investigates the role of taiko in Asian American culture through her experiences as a drum student in Southern California. In particular, she is interested in "the slippage between taiko as a specifically Japanese performance tradition, to its emergence as a Japanese American tradition, to its reformulation as a pan-Asian American tradition [as a Chinese-American Wong falls in this group], to its placement as a tradition open to any participants from any background."[28] At the time of writing she had studied for about two and a half years with her teacher, Rev. Tom, and considered herself barely intermediate. Like Kisliuk, Wong writes in detail about the experience of playing to encourage the reader to feel what she feels. There is the physical exertion of playing – the rapid heart rate, aching muscles, sweating, panting. There is the feeling of being part of a whole, and how the group members take care of each other. And on a deeper level there is personal transformation: "When I perform, I am someone else. My body changes automatically."[29] Part of that transformation involves shouting, an integral aspect of performance practice: "You aren't used to being loud, especially if you're Asian American."[30]

Being a long-time student afforded Wong greater access to her teacher than she would have had as an outside researcher. When Rev. Tom formed a group to perform publicly, Wong discovered that he was more interested in selecting students "who have the right attitude as much as anything else [. . .] it's not a matter of skill in itself."[31] Although Rev. Tom saw taiko as "a way to improve your life,"[32] he preferred not to preach Buddhist philosophy but allowed his students to discover it on their own. As Wong began to perform with other taiko groups, she was able to identify overlapping attitudes toward learning, repertory (which was connected to a politics of lineage involving anonymous versus individual composition), aesthetics, and poetics that allowed her to formulate a relatively specific encapsulation of taiko in America, despite a lack of codification.

Taiko performance drew Wong into a "sensual sounded body" and "noisy historical constructions" that asserted new presences. "The complications and the risks" of performing taiko, she concludes, "are so fundamentally part of it all that I must end by arguing they are intrinsically part of the pleasure – the pleasure of listening to taiko, of learning it, of performing it, of teaching it."[33]

Cultural competence and embodiment

Central to Katherine Hagedorn's research on the performance of Afro-Cuban *Santería* is the theory of cultural competence, derived from Chomsky's competence–performance dichotomy.[34] The theory of cultural competence maintains that one's knowledge of a tradition affects one's response to it (how the music means and inhabits the listener). Hagedorn identified at least four levels of competence among audience members at a performance by *Conjunto Folklórico Nacional de Cuba*: young foreign tourists, older Cuban religious practitioners (*creyentes*), young Cubans involved with the black market, and young dancers/singers who aspired to join the troupe (*aficionados*). Each of these interacted with performance in different ways and to different degrees, thus shaping the dynamics of the event.[35] Hagedorn began initially as a tourist, but through field research and practical study on the batá drums moved into other categories:

> During my classes at the FolkCuba workshop, my performative intent leaned toward that of an *aficionada,* as I excitedly memorized the rhythmic phrases and dance steps of 'the *orichas*.' [...] During the most recent years of my fieldwork, I have found myself sitting with the *creyentes,* singing along with the performers and swaying back and forth.[36]

Hagedorn's own drumming, and the fact that she ultimately became a practitioner of *Santería,* increased her cultural competence and deepened her reaction to folkloric events to the point that hers became an embodied knowledge, rooted in the physical as well as the cognitive planes.

Identifying the "field"

In the foregoing examples, ethnographers enter the field and then learn to perform. But sometimes it is performing that leads ethnographers to the field. This was the case with Mirjana Laušević, a Bosnian from Sarajevo who did her graduate studies at Wesleyan and subsequently taught at the University of Minnesota. As a Wesleyan student leading a group that performed songs of the former Yugoslavia, she stumbled upon the American Balkan scene – a virtual community of diverse Americans united by their love of "Balkan" music and maintained by a network of summer camps, Internet discussion boards, newsletters, and local music events. "The field itself may not be visible," wrote Laušević in her resulting ethnography *Balkan Fascination,* "but the resultant patterns are clear."[37] Her skill as a performer gained her entry as a teacher at Balkan camps and workshops, but it was her ethnicity that bestowed cultural authority in the eyes of her students. At the beginning stages of fieldwork she wrote, "I was not perceived as a 'scholar' who was there to learn, but rather as a 'native' who must be there to teach, criticize, judge, and evaluate."[38] As a "Balkan," she was expected to perform Greek dances and various other "native" traditions outside her experience. Laušević eventually negotiated these contradictions, and performance became one important means by which she skillfully identified insiders and outsiders in a "virtual village" that was at once vague and specific.

Challenges of performance as research

The role of performance in research is shaped by academic training and the values of institutions; as long as ethnomusicologists are valued as scholars first and performers second (if at all), performance is likely to remain an ancillary strategy.[39] Some anthropologists who use music performance as a lens for studying cultural issues resist performance as a research strategy on ideological grounds. For example, Michelle Bigenho, who works on Bolivian national and ethnic identities, believes that privileging "'doing music' over other kinds of fieldwork presentation plays into Western ideologies about music, talent, giftedness, and so on – all points that should be under anthropological scrutiny rather than assumed as givens."[40] But this is a minority opinion among ethnomusicologists, who tend to use performance either as a point of departure or to complement other research methodologies. Those who avoid or limit performance as a research strategy do so for much more practical reasons.

Despite Seeger's high degree of involvement in all aspects of Suyá culture, for example, he declined to be included in a pre-existing kinship network because of factional differences; he solved the problem by adopting an Indian "sister for ceremonies" who was also an outsider.[41] Hagedorn studied the male domain of batá drumming, knowing that she would never learn all of its secrets because of her gender and nationality.[42] Time is also a factor. The luxury of spending two years in the field, as Seeger and Kisliuk did, is increasingly rare in this era of shrinking financial support; consequently, performance opportunities may be less numerous. Age, marital status, ethnicity, and religious background, among other factors, may prevent a researcher from participating as a performer in certain circumstances.

Participation in performance can be revelatory, as long as one keeps in mind what is being revealed. For example, Hagedorn's drumming may sound Cuban and Rice's *gaida* playing may sound Bulgarian, but they are performing with their own consciousness; they do not share the lived experiences and psychologies of their collaborators.[43] Jeff Titon addresses this point when he writes that despite his enthusiastic participation in the singing of Old Regular Baptists, he never became one of them. Asked how he could sing with them without believing as they do, he replied, "I think I believe in music as they do"[44] – a statement that all those who employ performance as research in ethnomusicology would likely endorse.

Notes

1. As this chapter went to press, the second edition of Gregory F. Barz and Timothy J. Cooley, eds, *Shadows in the Field: New Perspectives for Fieldwork in Ethnomusicology* (Oxford University Press) was announced. For more on PAR, *see* chapters by Rice, Wong, Kippen, Shelemay, and Kisliuk.
2. For an overview of specific contributions from these fields, *see* Gerard Béhague, "Introduction," in *Performance Practice: Ethnomusicological Perspectives*, ed. Gerard Béhague (Westport, CT: Greenwood Press, 1984); Ruth Stone, *Theory for Ethnomusicology* (Upper Saddle River, NJ: Pearson/Prentice Hall, 2008).
3. Carol Silverman, "Learning to Perform, Performing to Learn," *Journal of American Folklore* 108, no. 429 (1995): 309.
4. Alan P. Merriam, *The Anthropology of Music* (Evanston, IL: Northwestern University Press, 1964).
5. Mantle Hood, "The Challenge of 'Bi-Musicality,'" *Ethnomusicology* 4, no. 2 (1960). See Spiller, this volume, and John Baily, "Ethnomusicology, Intermusability, and Performance Practice," in *The New (Ethno)Musicologies*, ed. Henry Stobart (Lanhan, MD: Scarecrow Press, 2008).
6. John Blacking, "Field Work in African Music," in *Reflections in Afro-American Music*, ed. Dominique-René de Lerma (Kent, OH: Kent State University Press, 1973), 214.
7. Quoted in Keith Howard, "John Blacking: An Interview," *Ethnomusicology* 35, no. 1 (1991): 60.
8. Quoted in ibid.: 61.
9. *See* Blacking, "Field Work in African Music," 208–9; Howard, "John Blacking: An Interview," 67.
10. Marcia Herndon, "The Cherokee Ballgame: An Ethnomusicologist's View," *Ethnomusicology* 15, no. 3 (1971): 340.
11. Milton Singer, "The Cultural Pattern of India," *Far Eastern Quarterly* 15 (1955).
12. Marcia Herndon and Norma McLeod, eds, *The Ethnography of Musical Performance* (Norwood, PA: Norwood Editions, 1980).
13. *See* Anthony Seeger, *Why Suyá Sing: A Musical Anthropology of an Amazonian People* (Cambridge: Cambridge University Press, 1987), 138–9.
14. Ibid., 115.
15. Ibid., 120. Kisliuk describes a similarly instructive breach of etiquette; *see* Michelle Kisliuk, *Seize the Dance: Baaka Musical Life and the Ethnography of Performance* (Oxford: Oxford University Press, 1998), 105.
16. Seeger, *Suyá*, 123.
17. Ibid., 86.
18. Ibid.
19. Ibid., 82, 120.
20. Timothy Rice, *May It Fill Your Soul: Experiencing Bulgarian Music* (Chicago: University of Chicago Press, 1994), 98.
21. Ibid.

22. Ibid., 102.
23. Ibid., 114.
24. Kisliuk, *Seize the Dance*, 101.
25. Ibid.
26. Ibid., 26.
27. Deborah Wong, *Speak It Louder: Asian Americans Making Music* (New York and London: Routledge, 2004), 229.
28. Ibid., 195.
29. Ibid., 196.
30. Ibid., 202.
31. Ibid., 207.
32. Ibid., 199.
33. Ibid., 229.
34. Katherine J. Hagedorn, *Divine Utterances: The Performance of Afro-Cuban Santeria* (Washington, DC: Smithsonian Institution Press, 2001), 57–8.
35. Ibid., 63.
36. Ibid., 64.
37. Mirjana Laušević, *Balkan Fascination: Creating an Alternative Music Culture in America* (Oxford: Oxford University Press, 2007), 13.
38. Ibid., 7.
39. *See* Silverman, "Learning to Perform, Performing to Learn," 307.
40. Michelle Bigenho, "Why I'm Not an Ethnomusicologist: A View from Anthropology," in *The New (Ethno)Musicologies*, ed. Henry Stobart (Lanham, MD: Scarecrow Press, 2008), 28–9.
41. Seeger, *Suyá*, 122–3.
42. Hagedorn, *Divine Utterances*, 89–90.
43. *See* Silverman, "Learning to Perform, Performing to Learn," 312.
44. Jeff Todd Titon, "Bi-Musicality as Metaphor," *Journal of American Folklore* 108, no. 429 (1995): 296.

13

The Three Configurations of Studio-Art PhDs[1]

James Elkins

It seems to me that the PhD in visual arts is inescapable: it is on the horizon. What is needed is an investigation into the conceptual shapes that the new degrees might take. The United States is well positioned to do this, and to become the place where such programs are rethought from the ground up. At least there's a chance of doing this in the United States, because it does not have the administrative structures (such as the RAE) and jargon (the call for a new definition of "research") that shape doctoral programs in the United Kingdom.

I propose three configurations that such degrees might take. My notion in this chapter is to sketch these models in a neutral fashion, as a kind of philosophic problem. The first model is relatively common, and the second and third are rare but, I think, preferable. Each model could be implemented in a number of ways: the student's research, for example, could be weighted as two-thirds of the degree, and the visual art as one-third. I leave aside all the pressing problems of the job market. I do not mention the fact that the new degrees have spread quickly in the United Kingdom in part because departments get funding based on how many advanced students they have – and that PhD students generate more money than MA students. (That, I think, is the elephant in the room in all UK discussions of the degree: they may sound disinterested, or motivated by ideas alone, but the raw fact is that institutions get money for PhD students, so it pays to establish these programs.) I don't raise the question of whether graduates with the new degrees would have an unfair advantage over those with MFAs. And I don't discuss whether most student artists at the MFA or MA level are capable of writing 50,000-word dissertations or doing PhD-level research.

In short, I pretend that the new degree has no economic, practical, or political dimensions. I do this because it seems very important to consider what the degree might mean for intellectual and creative life in the university. Can it contribute new ways of thinking about interdisciplinarity? Can it help reconfigure the conventional ways of conceptualizing the difference between making something and studying it? Can it help justify the presence of studio art departments in universities? Can it provide models for bridging history, theory, criticism, and practice – models that might have meaning even beyond the humanities?

None of this is to say that the philosophic issues can help solve the economic, practical, or political problems – or that those problems are less important. But it would be a pity, I think, to see the new creative-art PhD spread through the United States and Europe, and not be theorized as cleanly as possible. What is missing is a theorization of the possibilities in the abstract, before the exigencies of actual departments and resources come into play.

First model: the dissertation is research that informs the art practice

The most obvious relation between the PhD candidate's scholarship – the written dissertation or thesis – and her creative work is that the dissertation informs the artwork. The artist positions her scholarship so that it variously supports, modifies, guides, or enables her art practice. Within this first large grouping I distinguish five kinds of written dissertations, depending on which department supervises the project.

1 The dissertation is art history. The most common option is an art historical dissertation on the practices that lead up to the writer's own approach. The student would normally have a supervisor in art history, and one in studio art. In some Australian theses, the dissertation is written in anthropology, archaeology, sociology, or geography, but art history is most common. One advantage to this model is that, in theory, the candidate would be able to teach in a department of art history (or anthropology and so on) as well as in a studio department.

A guiding assumption of this configuration is that art history can strengthen, or at least productively inform, art practice. I think the point is often true, but it is debatable as an assumption, if only because so many artists have done so well by misinterpreting, travestying, simplifying, or otherwise distorting works and ideas that an art historian might say are most pertinent to their work. It's also the case that moments in art history were made possible by the artist's carelessness or ignorance of the relevant art history. It can be argued that German Expressionism depended on an insouciant disregard of academic criteria.[1] If Kirchner or Nolde had acquired PhDs in the history of German art, it is possible that they might never have been able to break the grip of academic work as effectively as they did. It matters that there have been times and places in art history where it would have been inappropriate to educate artists using a theoretically intensive regimen of research and writing. In some cases such an education might even have hurt the resulting practice.

2 The dissertation is philosophy or art theory. An artist's scholarship can also support her practice if the scholarly component is philosophy rather than art history. The dissertation might be a philosophic investigation of, say, the phenomenology of video practice instead of the history of video. A philosophic thesis, in this context, can be thought of as a professional outgrowth of the ubiquitous artist's statement. It could be supervised in a philosophy department, or in an art department, or art history: but the supervisor would, in this case, be treating the dissertation *as philosophy or theory* rather than history.

The same questions of relevance apply here: even though the PhD student might believe her practice is supported by her philosophic inquiry, the relation might appear very differently to her viewers, critics, and historians. Often artists' theories turn out to be irrelevant to what comes to be taken as most important about the work. And as studio instructors know, students who construct elaborate theories about their work sometimes use theory not for its content as much as its rhetorical force: the philosophy or theory of art serves as a smokescreen, hiding what is actually of interest in the work.

3 The dissertation is art criticism. The student's scholarship can also support her art practice if the dissertation is art criticism rather than philosophy or art history. This is, I think, the most common form of creative-arts PhD dissertation, and it can also be found outside the visual arts. I have seen an example produced in the PhD program in creative writing at the University of Houston. The student, Mark Caughey, wrote a 150-page critical examination of his own poetry, which served as an introduction to the dissertation (which was a book of his poetry).

In the United Kingdom, creative-art doctorates of various kinds have been around long enough to get an uneven reputation; some are not much more than over-extended Master's theses, with a written component that is essentially critical in nature, with an admixture of art history and art theory. They tend to be supervised in art history or in the relevant art department. They are, in that sense, very similar to the theses ("dissertations" in the UK and Ireland) that are written by MA or MFA art students; those texts tend to be mainly art criticism, aimed at elucidating the student's practice. The most immediate challenge to the development of the studio-art PhD in the United States is to find ways of preventing it from slumping into a protracted MFA thesis. To that end, it is important to reconsider two issues that are constitutive of art criticism in the academy.

The first is self-reflexivity. The purpose of the juxtaposition of art criticism and artwork at the doctoral level would presumably be to reach a pitch of sophistication in the description and evaluation of one's own art, on the reasonable grounding assumption that improving self-reflexivity is a central purpose of graduate study. As far as I know, self-reflexivity is not doubted as a goal in any graduate studio art program.

Yet many artists have made compelling work even though they had no idea of the critical matrix to which their work belongs, and despite the fact that they were only minimally reflective about their own practice. It is also true that some artists' work thrives on self-awareness; for artists of that kind the new PhD degree might be ideal – although there is no account of what kinds of art have been best served by self-awareness. This idea that self-awareness is a desideratum for PhD-level instruction needs to be treated as a problematic assumption, not a guiding principle.

A second issue with creative art dissertations in the form of art criticism is that the subject of art criticism is virtually never taught in PhD programs in philosophy or art history.[3] Art criticism appears as a historical subject in art history curricula – there are courses on Baudelaire, Diderot, and so forth – but not as a practical subject. In the absence of structured sequences of courses on practical art criticism, it is dubious that art-critical dissertations can be effectively read and critiqued *on a PhD level*. It would, of course, be possible to find philosophers or art historians who could assess such dissertations, but only for their logic (if they were read as philosophy) or historical veracity (if they were judged by art historians).

4 The dissertation is natural history, or economics, or any number of fields outside the humanities. In this option, the candidate looks further afield than art, art history, or philosophy. At a 2003 session of a conference of art schools in Los Angeles (NASAD/AICAD), David Williams, Chair of Art at the Australian National University in Canberra, said that it is a very popular option among his students to write a "subthesis" in the sciences. The student has an art practice, in any medium, and chooses to obtain a PhD in biology, say, or in genetics: whatever field they are qualified to enter.

5 The dissertation is a technical report. There are media and kinds of art making that are not fully mastered by students at the MFA level. At Alfred, New York, a school well known for its ceramics program, there is a laboratory that specializes in high-tech, non-art ceramics (at some point they made the tiles that protected the Space Shuttle). That laboratory is not utilized by the MFA students as much as it could be because the students lack the education in inorganic chemistry. A PhD program in ceramics could remedy that. Print-making techniques like metal engraving or certain advanced fabric techniques are commonly omitted from the MFA; they could be taught given a few more years' worth of courses.

There are many advanced industrial materials that are not taught in art schools, and a PhD would be a way to institute a program in the relevant contemporary materials science. Such a degree would also help meliorate the disjunction between current engineering, with its many sophisticated materials, and art practice, which still keeps mostly to oil, clay, metals, paper, and wood.[4]

In a sense this kind of PhD exists in a few North American and UK institutions, where students continue their studio practice for several years beyond the MFA, eventually earning the PhD (or DLitt). However those programs are not PhDs in the sense I am speaking of here, because they do not combine ongoing studio practice with scholarly work at the PhD level.

Second model: the dissertation is equal to the artwork

In the first model, the dissertation is a repository of research that informs or otherwise aids the art practice. In the remaining models, the dissertation is implicated in the artwork, or even considered *as* the artwork. That has the advantage of freeing the scholarship from its ultimately informational or supportive role, and potentially making the research equal to the artwork – or even making it *into* the artwork. In this second model, the dissertation is considered as conceptually equal to the art. The research doesn't support or inform the art, but complements it, each one illuminating the other. Rather, the relationship between research and artwork can be divided into two possibilities.

1 Research and artwork comprise a new interdisciplinary field. In this case the creative-art PhD might be considered as an example of the confluence of disciplines that is currently congealing into the field called visual studies or visual culture. The new PhD becomes an opportunity for a student to collect an idiosyncratic collection of disciplines, with art just one equal among others in a collection of disciplines. It differs from the example above (in which the student writes a dissertation in a non-art subject) because there, the non-art field is used to inform the art project; here, the non-art field or fields are all taken to be equal contributors to a new constellation of interests.

2 Research and artwork are understood as wholly separate projects. This possibility is like the previous one, in that the student's art practice and the associated non-art disciplines are imagined to be equal participants in the PhD process. The difference is that the new configuration of fields is not understood as a potentially coherent project, but as a juxtaposition, whose rationale does not need to be analyzed. Even the candidate herself might not be sure of the pertinence of her research interests; she might just have a strong interest in both video art, for example, and scholarship in some other field. In this case the function of the faculty would be to help advise the scholarly portion of the dissertation and the art practice at an appropriate level, leaving it up to the student to work through the possible connections between them. This is in some measure the model adopted by the Canberra School of Art.

I find this option the second-most intriguing. It is interesting to contemplate what an artist might accomplish by keeping two sets of activities separate from one another for the duration of a PhD program, without being asked to formally theorize their connection. It also seems wholly in keeping with the way art is often produced, in the company of many disparate interests that do not, at least for some time, seem to be directly linked to one another. It avoids the usual academic demands of coherence, rationality, and intellectual synthesis – which again is appropriate for much of visual art.

The models I have named up to this point have serious, if also potentially productive, conceptual flaws. The third and last model is the most interesting to me – and raises, potentially, even more difficult philosophic and practical problems.

Third model: the dissertation is the artwork, and vice versa

The final option that occurs to me is to imagine the scholarly portion of the thesis inextricably fused with the creative, so that the artwork is scholarly and the scholarship is creative. I have seen attempts at this, including a multimedia dissertation done at the University of Chicago that includes a CD, photo exhibit, 16mm films, and written dissertation. But I haven't seen examples in which the scholarship melts into creative work. In the University of Chicago thesis, for example, the writing is clearly situated in art history, anthropology, and film studies, and it remains distinct from the student's films, CDs, and photographs. But why not try to write a PhD dissertation in art history *as if it were fiction?*

I divide this into two possibilities:

1 The research dissertation is intended to be read as art, *and the visual practice* as research. The models for a more radical fusion of text into creative work would be writing by scholars such as Michel Serres or John Berger (who mingles poems and art history); there are very few such examples. If such a dissertation existed, it would be extremely difficult to evaluate in an academic setting because the entire apparatus of scholarship, from the argument to the footnotes, would have to be legible as creative writing.

2 There is no research component: the visual art practice, together with its exhibition and supporting materials, simply is *the PhD.* This last possibility is, perhaps unexpectedly, more common than the previous one. It has been practiced in a handful of institutions in the United Kingdom. The argument is basically that visual art practice should not borrow from other academic fields, but remain true to its own media and purposes. It has also been said that the creative-art PhD, in any of the forms I have been listing, is inherently unfair because it requires a student to complete doctoral level work in an academic field and also create doctorate-level visual art.

I think this last and most radical possibility is also the most interesting. It is a logical endpoint for the new degree, because each of the foregoing models presupposes that visual art practice can be taken to the level of the doctorate. This last option is simply more consistent than the previous models, because it permits the visual art practice to carry the burden of competence that will allow it to be taken as a doctoral-level accomplishment *aside from* whatever writing might support or augment it.

Of course, this final possibility presents severe problems when it comes to assessment. How is a studio-art instructor to determine if the studio practice is at PhD level? I think this question is, in its very form, unanswerable, and it may not be a productive approach to the problem of assessing the new degrees. It may be more sensible to ask, first, how supervisors might read and respond to dissertations that are produced in all the possibilities I have discussed *except* this final one. When guidelines for assessing those models are in place, it might make more sense to try to say what PhD-level assessment in creative art might look like.

Notes

1. This chapter is a revised and expanded version of a paper given at the annual meetings of the National Association of Schools of Art and Design (NASAD) and the Association of Independent Colleges of Art and Design (AICAD) in 2003. Earlier versions were published as "Theoretical Remarks on Combined Creative and Scholarly PhD Degrees in the Visual Arts," *Journal of Aesthetic Education* 38 no. 4 (2004): 22–31; *The New PhD in Studio Art*, no. 4 in the occasional series called *Printed Project* (Dublin: Sculptor's Society of Ireland, 2005); and as a précis titled "Ten Reasons to Mistrust the New PhD in Studio Art," *Art in America* (May 2007): 108–9. An in depth version is forthcoming in a collection on the new PhDs in studio art. An in-depth version appears in *Artists with PhDDegrees: On the New Doctoral Degree in Studio Art* ed. James Elkins (New York: New Academia Publishing, 2009).
2. James Elkins, "On Modern Impatience," *Kritische Berichte* 3 (1991).
3. Ibid., *What Happened to Art Criticism?* (Chicago: Prickly Paradigm Press, 2003).
4. Film, computers, and video don't count as advanced materials in this sense: they are fully developed technologies that are used, but not normally understood at the level of materials and processes.

14
Approaching Knowledge, Research, Performance and the Arts

Arthur J. Sabatini

For the contributors and most readers of this collection, it would seem obvious that art and performance are interdependent with research and practice and that art is significant in the creation, discovery, production, and circulation of knowledge. Yet, historically and in diverse institutions and discourses, this is neither self-evident nor accepted. From the viewpoint of public awareness and for some academics, the arts and performance, particularly as they have evolved over the past century, are considered difficult to demonstrably isolate as objects or entities. Moreover, and regardless of the area of study, the question of what precisely constitutes research and how knowledge is defined, produced, and communicated is problematic. Today, paradigms, methods, systems of verification, and boundaries of knowledge are persistently under examination in every field and discourse, whether in the arts, humanities, sciences, social sciences, or in multiple, developing inter- and cross-disciplinary areas.

Research and the processes of the creation, production, and uses of knowledge are subjects that have been explored historically in philosophy, critical theory, and social thought. Current investigations have extended study to encompass knowledge and: economics, ideology, institutional structure, and cultural politics. In addition, for some decades, technologization and interculturalization have altered thinking on issues of the evaluation and representation of knowledge. More recently, findings and theorization in neuroscience and cognitive studies have added other dimensions to the concept of knowledge and human ways of knowing.

For artists, scholars in the arts, and those in universities and institutions with vested interests in artistic research and creativity, the emergence of different types of arts research (in the visual arts, music, dance, and so on), performance studies, practice as research (PaR) and performance as research (PAR) provoke a variety of questions. These range from outright theoretical and philosophical matters to re-examinations of the interrelationships among the arts, performance, and knowledge, as well as queries surrounding the specific character of arts research, methods, and evaluation. There are also innumerable practical, professional, financial, and cultural issues of interest.[1]

This chapter will focus on issues concerning the arts, research, performance and knowledge. It begins with a general historical overview of philosophical responses to art and epistemology, then traces twentieth/twenty-first-century reconfigurations of thinking about knowledge in the sciences, social sciences, and the arts. Following comments on the arts and performance research in and out of the university, the chapter concludes with a schematic review of selected key thinkers and texts that position practice as research (PaR) and performance as research (PAR) in relation to current discourses and trends.

In traditional Western philosophy, aesthetics has focused on the arts with an emphasis on: imitation (*mimesis*) and representation, concepts of beauty and pleasure, aesthetic experience, and the evaluation and definition of art. From the eighteenth century onward, philosophers attempted to account for aesthetic experience in terms of the imagination, feeling, sensuous perception, moral judgment, and cognitive processes. Kant, Hegel, and Schopenhauer, among others, endeavored to formulate comprehensive, integrated aesthetic and epistemological theories and propose universal principles. However, their work, as subsequently critiqued, failed to account for many factors concerning artistic and aesthetic experience, particularly across genres. Their ideas also required re-examination in the context of the variegated artforms that began to appear in the twentieth century. At best, through the nineteenth century, art was unquestionably affirmed as an essential product of the human imagination and expressive of emotion. However, art's significance was limited to being a purely subjective phenomenon, experienced by individuals (for example, readers, spectators) who respond with "detached" pleasure and determine their own understanding. Art could be neither universal nor verifiable; and contemplation of art was considered distinct from rational and ethical thinking.[2] In short, as Mark Johnson succinctly notes, this line of philosophical inquiry concluded that "because feeling is noncognitive, it cannot give rise to knowledge."[3] This dichotomy – between thinking/knowledge versus feeling/emotion, knowledge of the mind versus of the body, and knowledge as present in language through rationality but not through the body and its forms of knowing – endures and, in some discourses, underscores present debates.

A second, historically significant (and still prevalent) perspective on knowledge derives from the traditions of empiricism, positivism, and scientific thought in general as it pertains to the "hard" sciences (which almost exclusively refers to study of the natural world). Accordingly, in public perception, scientific knowledge is characterized by "objective" observation, measurement and the "quantitative" recording of verifiable facts. Thus, knowledge, ideally, results from "value-free" practices and methods of research, including hypothesis testing, data collection, and validation. Aesthetic knowledge, in this context, is dismissed as being too subjective and neither reliably quantifiable nor verifiable.

Undoubtedly, throughout the modern era, the prevailing conception of knowledge derives from the idealized model of the "hard" sciences, with its implicit "objectivity," and use of mathematical models and systems for quantitative research. However, throughout the twentieth century, the assumptions underlying scientific thought and the "culture of science" have been subject to inquiry and rigorous appraisal.[4] Self-critical scientists, analytic and anti-empirical philosophers, epistemologists, and logicians have questioned the principals, processes, and overall institutional practices of the sciences.[5] They have raised questions about the relationship between objects and theory making, experimentation, and the processes involved in all stages of the scientific method. For Robert Crease, laboratories and other sites of scientific experimentation can be understood as versions of theatre and performance.[6] The first laboratories, he notes, were constructed with seats for spectators, as are many today (and they also have video or computer screens) and most experiments, as in theatre, are representations of phenomena and practice. Postmodern thinkers, such as Bruno Latour and Donna Haraway, and critical theorists have problematized the epistemic claims of science, its social and cultural praxis, and accounts of its structures of experimentation.[7]

The primacy of scientific knowledge and its ways of knowing have further been challenged by explorations and findings in relation to human cognitive processes and communication. Within all academic fields and disciplines, there has been systematic development of numerous qualitative approaches to research. These include qualitative methods such as ethnography and narrative analysis and forms of observer/participant interactions and performance. It is also necessary to recognize that sustained varieties of qualitative research methods – in fields ranging from education and gender studies to arts research and practice – have convincingly provided evidence and argumentation for the acceptance of multiple forms and conceptions of knowledge. In some areas, this has also contributed to the development of hybrid methodologies that combine quantitative and qualitative methods; or, in the case of the arts and performance, research practices are employed that integrate long-established activities or incorporate newer technologies.[8] In short, there are varied, if not competing paradigms for defining and approaching research, knowledge, and ways of knowing.

Turning to research in the arts and performance, several distinctions must initially be made about research as it is evaluated in education and academia, and research as it pertains to artists outside of the university, currently as well as historically and across cultures. In the context of the university, research, or creative activity, is expected as a complement to teaching. Research is supported, conducted, and assessed as it contributes to the pursuit of knowledge and pedagogy, professional development and contracted responsibilities. The discovery or creation of new knowledge and the advancement, refinement, synthesis, or reinterpretation of knowledge are frequently cited as definitions of research. The criteria for success and recognition of knowledge or creative work are established according to evaluation by peers and university policy (which varies across disciplines and in different universities). Standards are set for publication or, for creative performance work, its reception and reviews in theatres, performance venues, galleries, on-line, and so on. The amount of funding support for research can also be a factor in evaluations.

Outside of the university, historically, an artist's research or processes are generally subsumed in the work itself, either in singular productions or over the course of a career. Research and practical and technical skills are, if acknowledged at all, presupposed as being necessary for the completed work. This is, of course, relative: critics, scholars, and other artists often attend to research and technical accomplishment. In different cultural contexts, these factors can be highly valued, as with dance or musical performances. If an artist chooses, examples of types of research may be offered, or valued, when work is published or presented. Research is assessed, if it is, when an artist's work is collected and studied, which is to say, an artist's notes or archives may not be considered research. Some artists – playwrights, choreographers, composers – develop highly specialized systems and techniques for research, producing, and training others. However, unlike the research in the sciences or social sciences, an artist's research and methods, though they may be highly formalized, as in actor training or dance, are intended to be distinctive in relation to their work and future versions of it, if not the demands of their socio-historical situation.

In some cases, artists' research can function as a contribution to knowledge and to others' work or cultural performance and production. For example, dance revolves around investigations and knowledge of the body; and there are choreographers who combine training in ballet with their own or other modern techniques. The same holds true for actors and musicians. Moreover, and this is a crucial point: unlike in the sciences, social sciences, and other fields, artists' research and methodology are only partially "a means to an end" in the creation or discovery of knowledge. In many cases, research processes and methods are inherently themselves an exemplum, a demonstration, a thematic, or social, cultural, or human signifier. Consider breathing. Breathing techniques, for dancers, actors, and musicians are generally based on ancient texts, which are modified, over centuries through current practice in any number of cultures. Obviously necessary to control the body and voice, they also vitally contribute to an actor's presence, expressivity, or spirituality.

Of course, while there might be exceptions, few artists outside of universities are formally "evaluated" on the basis of their research. The "success" or result of their research, as some of the chapters in this book demonstrate, is relative to their legacy and how their ideas, systems, and work continues or is adapted. To complicate the issue, while many artists have always been engaged in research, they may not have defined what they do as research (let alone as qualitative research), nor as a method for any purpose other than doing their own projects. Since, arguably, public reception of art and performance is focused on presentation, there is incidental concern for art's underlying sources or theory, stages of development and research.

To the degree that performance or practice-based research/performance as research and its variations are to be incorporated into university systems, it must first account for the vast range of existing artistic research and practice. It seems necessary to: (1) re-examine and historicize research by artists outside of academia and relate or adapt it to existing discourses and approaches to research (including scientific experimentation) in the university; (2) present the philosophical grounding for art and performance in relation to knowledge and ways of knowing; (3) draw upon examples of emerging research and qualitative methods in the arts and education, and demonstrate the connections to performance studies, PaR and PAR.

Conclusion

Regarding the history of research in the arts and performance, there have been and still remain deep, formidable, and wide-ranging traditions of artistic research and a multiplicity of methods for creating art and training artists and performers. It seems simplistic not to recognize that the training of performers is built upon considerable research into the capacities of the human body and mind. Use of the voice, breathing, manual dexterity, movement techniques, directing or choreographing for performance are all outcomes from highly proscribed and ever-evolving systems that have been researched, repeatedly tested, and advanced by practitioners worldwide. Similarly, visual and technologically oriented artists engage in variegated forms of research. Painters research chemicals and other materials in order to find precise pigments and textures. Research and experimentation by musicians in every culture and era has been conducted in order to explore sound, aurality, acoustics, instrument building, and performance techniques. Writers, before the Internet, traveled and produced ethnographies.

In certain periods and contexts, research has been an intentional program of artists or collectives. Russian Constructivists and Bauhaus artists identified their sites as laboratories and conscientiously developed programs for research, experimentation, and teaching. Similarly, the Surrealists established the Surrealist Bureau of Erotic Research. Another example, which consciously drew upon anthropological theory and techniques were the poets and performers associated with Ethnopoetics. The American Society for Theatre Research has defined agendas for its approaches and, as some chapters in this book demonstrate, there are legacies of research in theatre and performance by Stanislavsky, Grotowski, and others. Historical and many contemporary examples can be added to this list, more recently in relation to artforms incorporating technology and the information arts where quantification is routinely practiced.[9]

To approach questions of art, performance, and knowledge philosophically and critically, there are a range of discussions in diverse fields and discourses. Revisionist and emergent theories in aesthetics and epistemology have laid the groundwork for a reconsideration of art, knowledge, and the interrelationships among the arts. Jean-Luc Nancy reconsiders the idea of disciplinarization in *The Muses*; within analytic philosophy, Nelson Goodman argues that art and science function in distinctive symbolic systems, or as a language. He also addresses performance and distinguishes between what he calls autographic works (a painting or other such singular piece) and allographic work (such as a score or dramatic text). Goodman flatly states, "the arts must be taken no less seriously than the sciences as modes of discovery, creation and the enlargement of knowledge."[10] Other studies, such as Gérard Genette's *The Work of Art*, David Davies's *Art as Performance*, James Young's *Art and Knowledge*, and Richard Shusterman's *Performing Live: Aesthetic Alternatives for the Ends of Art* provide various cognitivist, epistemological, and contextualist approaches to rethinking art and knowledge. Significantly, while these studies derive from a tradition of aesthetic philosophy, they encompass performance, as both an active concept and in reference to theatre.

Human knowledge, cognition and the body, and versions of embodiment have been long recognized in phenomenological thought, the pragmatism of John Dewey, and in performance studies and performer training. Contesting mind/body dualism, investigations in cognitive studies and neuroscience have provided artists with a richer perspective on their practice and work.[11] Other developments appear in Mark Johnson's *The Meaning of the Body: Aesthetics of Human Understanding*. Johnson demonstrates that findings in neuroscience affirm the more philosophical insights of Dewey and Merleau-Ponty's phenomenology of the body. Dewey's work deeply informs educational theory (as well as the philosophy of Goodman) and scholars like Elliot Eisner and Graeme Sullivan elaborate on Dewey's thought. Sullivan's *Art Practice as Research: Inquiry in the Visual Arts* discusses a number of art projects that result in performance.[12]

There is a growing body of inquiry on arts research in education. In addition to texts mentioned in this volume's introduction, Knowles and Cole's *Handbook of the Arts in Qualitative Research* brings together essays that address methodologies, research genres (including performance), inquiry processes, and interdisciplinary perspectives.[13] In relation to pedagogy, it has been well established that this decidedly falls into a type of qualitative research.[14]

Finally, in the context of performance studies, critical theory, and postmodern thought, there are extended approaches that argue for performance as an epistemological category and tool. "Emphatically," Theodor W. Adorno asserts, "art is knowledge, though not the knowledge of objects. Only he understands an artwork who grasps it as a complex nexus of truth, which inevitably involves its relation to untruth, its own as well as that external to it; any other judgment of artworks would remain arbitrary."[15] Adorno writes chiefly about music and literature; however, his approach to art and epistemology, particularly in *Aesthetic Theory*, is directly applicable to dance, theatre and performance. Regarding performance, Jon McKenzie, among others, argues that it is a near ubiquitous term affecting every realm of contemporary life. His book traces performance studies as a discipline, but also performance in science and business. For McKenzie, "knowledge on the performance stratum, then, is characterized by simulation rather than representation, by the instability of subjects and objects, and by the playing and contesting of diverse language games."[16]

It is reasonable to say that this is a moment of discovery and transition where the arts, research, and performance are realigning and providing a richer understanding of knowledge and how we know. I suspect that aesthetic research and performance will prove to be increasingly significant to human knowledge as more concentration and study occurs.

Notes

1. *See* Introduction for distinctions among performance studies, PaR, PAR, and so on; see also Jon McKenzie, *Perform or Else: From Discipline to Performance* (London and New York: Routledge, 2001).
2. For more on Kant et al., *see* Lucien Krukowski, *Aesthetic Legacies* (Philadelphia: Temple University Press, 1992). *Re* aesthetics and ideology, *see* Terry Eagleton, *The Ideology of the Aesthetic* (London: Blackwell, 1990).
3. Mark Johnson, *The Meaning of the Body: Aesthetics of Human Understanding* (Chicago: University of Chicago Press, 2007), 218.
4. Andrew Pickering, *Science as Practice and Culture* (Chicago: University of Chicago Press, 1992).
5. *See* A. J. Ayer, *Language, Truth and Logic* (New York: Courier Dove Publications, 1952); Paul Feyerabend, *Against Method* (London and New York: Verso, 1993).
6. Robert Crease, *The Play of Nature: Experimentation and Performance* (Bloomington: Indiana University Press, 1993).
7. David Gooding, Trevor Pinch, and Simon Schaffer eds, *The Uses of Experiment: Studies in the Natural Sciences* (Cambridge and New York: Cambridge University Press, 1989).
8. Nagy Hesse-Biber and Patricia Leavy eds, *Approaches to Qualitative Research: A Reader on Theory and Practice* (New York and Oxford: Oxford University Press, 2004); J. Gary Knowles and Ardra L. Cole, eds, *Handbook of the Arts in Qualitative Research* (Los Angeles: Sage Publications, 2008).
9. Stephen Wilson, *Information Arts: Intersections of Art, Science, and Technology* (Cambridge, MA, and London: MIT Press, 2002); Steve Dixon, *Digital Performance: A History of New Media in Theater, Dance, Performance Art, and Installation* (Cambridge, MA: MIT Press, 2007).
10. Nelson Goodman, *Languages of Art: An Approach to a Theory of Symbols* (Indianapolis: Bobbs-Merrill, 1968), 102.

11. *See* Bruce McConachie and F. Elizabeth Hart, eds, *Performance and Cognition: Theatre Studies and the Cognitive Turn* (London and New York: Routledge, 2006).
12. A growing number of books examine various forms of practice as research or "imaginative" research. *See* E. Barrett and B. Bolt, eds, *Practice as Research: Context, Method, Knowledge* (I. B. Tauris, 2007); I. Edgar, *A Guide to Imagination-Based Research Methods* (New York and London: Routledge, 2004). *Also see* Graeme Sullivan, *Art Practice as Research: Inquiry in the Visual Arts* (Thousand Oaks, CA: Sage Publications, 2005).
13. Knowles and Cole, eds., *Handbook of the Arts*.
14. In the field of education, arts-based research is internal to pedagogy and not necessarily for the training of future artists: "The central purpose of arts-informed research is knowledge advancement through research, not the production of fine art works. Art is a medium through which research purposes are achieved. The quality of the artistic elements of an arts-informed research project is defined by how well the artistic process and form serve research goals," from the chapter "Arts-Informed Research" in Knowles and Cole, eds, *Handbook of the Arts*.
15. Theodor W. Adorno, *Aesthetic Theory*, ed. Gretel Adorno and Rolf Tiedemann, trans. C. Lenhardt (London and Boston: Routledge & Kegan Paul, 1984), 262.
16. McKenzie, *Perform or Else: From Discipline to Performance*, 180. Relevant texts noted in this essay, but not cited include: David Davies, *Art as Performance* (Malden, MA, and Oxford: Blackwell, 2004); Elliot Eisner, *The Arts and the Creation of Mind* (New Haven: Yale University Press, 2002); Gérard Genette, *The Work of Art: Immanence and Transcendence*, trans. G. M. Goshgarian (Ithaca, NY, and London: Cornell University Press, 1997); Jean-Luc Nancy, *The Muses* (Stanford, CA, Stanford University Press, 1996); Richard Shusterman, *Performing Live: Aesthetic Alternatives for the Ends of Art* (Ithaca, NY, and London: Cornell University Press, 2000); James O. Young, *Art and Knowledge* (New York: Routledge, 2001).

Part II

Cartographies – Terms for Finding/Charting the Way(s)

15
Action Research

Kim Yasuda

Action research (AR) – an emergent term for humanities and arts-based practices in the US academy for little more than a decade – has influenced research and knowledge creation in the fields of social and educational sciences since the mid-1940s elsewhere.[1] First cited in the work of social psychologist Kurt Lewin, AR was framed in methodological terms as an iterative process, coupling a body of research with its outcome through a spiraling cycle of planning, action, and fact-finding.[2]

The union of *action* + *research* implies a level of "activation" in the acquisition of knowledge – an inherent sense of experimentation with new forms of knowledge production. The traditional separation between theory and practice moves toward new forms of critical engagement that are not necessarily singular or focused in pursuit, but are complex, public, and collaborative in nature, providing a level of social engagement within the investigative process that influences and is influenced by a potential set of outcomes.

AR in the arts and the academy

AR has evolved in arts practices over the past decade and has been framed as such by its academic enterprise – a system that invests in the critical relationship between theoretical and practical investigation as part of a creative practice (for example, the emergence of theory requirements within studio art programs). Lewin's cyclical form of investigation as a process might very well resonate with the creative impulses of the artist; however, in humanities and arts "scholarship" within the academy, neither the value of AR nor its complex politics have been completely sorted out. For example, AR has primarily been engaged in the fine arts in order to legitimize the value of the practicing arts as a scholarly form of research in parity with other fields of academic investigation. Questions such as, "Is art research?" or "When is art research?" would not be a set of propositions in other disciplines where there is inherent qualification of the research as such.

While less so in the performing arts, which may be more collaborative in nature, scholarly research in the humanities and visual arts has been institutionally rooted and valued in terms of singular or studio authorship (the monograph or solo exhibition). The term "action research" has been appropriated from within this academic context, implying a new set of propositions for the humanities and arts scholars to consider in pursuit of an alternative research paradigm that challenges those authorship models with configurations of collectivity and public agency. This linked notion of *action-research* requires a level of engaged scholarship that conflates the division of labor within a scholarly profile in the order of *research* as is inextricably connected *to teaching* and *service* over its hierarchical antecedent. However, academic institutions have yet to reckon a current value and rewards system with the open-ended field of conditions implied in AR as it migrates from the social sciences into the humanities and arts.

AR potentialities

Outside of the academy, the history of arts practice throughout the twentieth century could be said to have engaged in AR methods, whether considered research or not. Since the European avant-garde, the visual and performing arts have effectively explored imaginative forms of activated research from situational and relational aesthetics to social experiments in collectivist and interventionist actions, giving process and reception equal weight with the production of forms. For example, Joseph Beuys used creative research to actively participate in politics and social change; he developed the idea of social sculpture and founded the FIU: The Free International University of Creativity and Interdisciplinary Research.[3]

Within the academy, contemporary forms of AR specific to both teaching and research in the arts have begun to explore the capacity to take on this larger and more complex set of propositions – as a form of social and pedagogical transformation of the definition and formation of university scholarship. The movement to embrace these variant forms of public scholarship in the arts appears to be taking place in a number of ways: first, as case-study innovations within and outside the university classroom on the part of a creative faculty in collaboration with their students, and second, as a national coalition of university administrators and faculty who advocate for radical change in the current university research/reward system toward the more democratic model of research as implied by the linking of *action* and *research*.[4]

In both cases, there is a collective force at work that is challenging the boundary-based disciplinary practices we have known since the modern institutionalization of educating artists, which now must effectively respond to the rapidly changing cultural environments of the twenty-first century. AR is a potential vehicle for a paradigm-shift in higher education with regard to disciplinary research and its relationship to practice. AR and the risk-taking innovation implied in its practice-linked research is one of the few forms that effectively breaks away from traditional models to become a key catalyst in new scholarly formations.

Notes

1. There is a large body of work on AR; most recently, *see* Mary Brydon-Miller, Patricia Maguire, and Alice McIntyre, eds., *Traveling Companions: Feminism, Teaching, and Action Research* (Westport, CT: Praeger Publishers, 2004); Davydd Greenwood and Morten Levin, *An Introduction to Action Research: Social Research for Social Change* (London and Thousand Oaks, CA: Sage Publications, 2007); Peter Reason and Hilary Bradbury, eds, *Handbook of Action Research: Participative Inquiry and Practice* (London and Thousand Oaks, CA: Sage Publications, 2008); Ernest T. Stringer, *Action Research*, 3rd edn (London and Thousand Oaks, CA: Sage Publications, 2007). For a practical guide, *see* Jean McNiff, Pamela Lomax, and Jack Whitehead, *You and Your Action Research Project* (London and New York: Routledge, 1996).
2. Lewin, "Action Research and Minority Problems," in Greenwood and Lewin, *An Introduction to Action Research.*
3. Collectivist practitioners such as Guy Debord and the artists of the Situationist International explored the AR relationship between forms of critical writing and visual production. In the USA, for example, artists Gordon Matta-Clark and Donald Judd, while recognized in the art world for their formal production, actively engaged their work in a social practice through their writing, activism, and public interventions.
4. The research university consortium, Imagining America: Artists and Scholars in Public Life has been most effective in its efforts to encourage fundamental institutional change through the advancement of public scholarship modalities such as AR. Established in 1999 IA's inaugural White House conference produced collective support for a community comprised of university presidents, faculty artists, scholars, and community cultural leaders. This consortium argues: (1) for the importance of civic work within a creative practice; (2) that campus–community cultural partnerships be recognized by colleges and universities as innovative; and (3) for the legitimacy of such practices within the professional economy of the university. A crucial feature of IA's mission is to advocate for systemic change within the academy and for the valuing of AR practices within the system of tenure and promotion of faculty. Its advocacy platforms include annual conferences; a "tenure team initiative"; the collaborative development of policy documents to be circulated among university research review committees across the nation; a series of scholarly publications that encourage discourse on Action Research practices; and a consortium membership exchange between institutions to facilitate the emergence of new learning environments. *See* www.imaginingamerica.org.

16

Disjunction: Performing Media Space

Nick Kaye

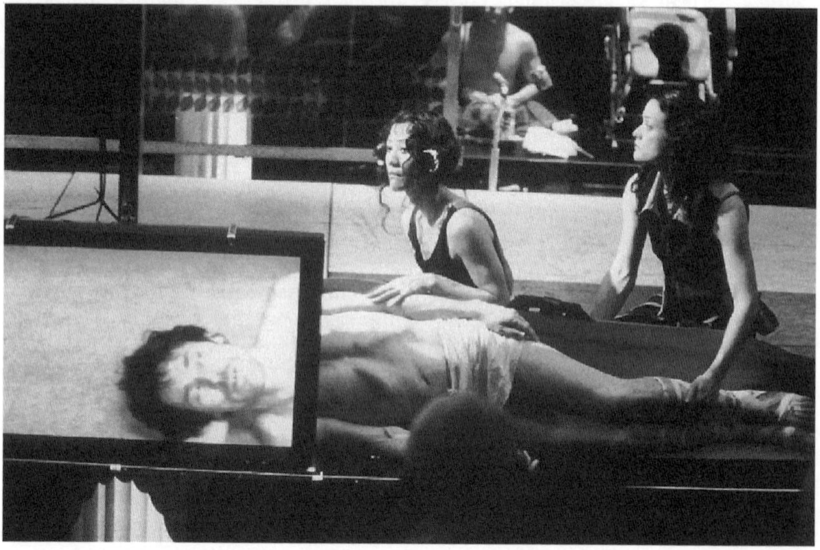

Figure 6 *To You, The Birdie! (Phèdre)*, directed by Elizabeth LeCompte. Pictured (left–right): Willem Dafoe, Koosil-ja Hwang, Ari Flakos, and Dominique Bousquet. Photograph by Mary Gearhart

"Media space," the artist Tony Oursler has suggested, "is a conglomerate of virtual spaces"[1] – a space whose boundaries and identities are multiple, permeable, and contradictory. Like the electronic "mediation" of "real" times and spaces, "media space" presents a collocation of spaces and states in paradoxical or dissonant relationship. Multi-media performance, in its theatrical enactment and realization of "media space," tends to amplify these instabilities, even where its surfaces may suggest otherwise. In the multi-media theatre of the Wooster Group, The Builders Association, or John Jesurun, mediated images are deployed in seemingly paradoxical ways: to evidence "the real" performer or the "real act"; to amplify and articulate the "live"; to recover and modulate "performer presence."

In the performance of media spaces, the practices of video art, performance art, and multi-media theatre frequently intersect or mirror one another – an extension of the "overlappings" defining media space. Oursler's work interrogates this "media space" in relocations of projected or screened images onto three-dimensional objects – including dummies, mannequins, flowers, and abstract forms – to produce uncanny, seemingly "interactive" video "entities" in alarming collocations of "real" and "virtual" materials and spaces. For example, *Crying Doll* (1993) combines the projection of a perpetual act of crying onto the "face" of a doll-size dummy to elicit sentimental identification on behalf of the viewer.[2] Yet, like many of his installations, *Crying Doll* exposes the props and mechanisms of its operation: the untransformed materials comprising its seemingly unfinished body; its simple wooden support; the projector focused on its face, as if surveilling rather than animating its activity. Here, the very elements that assert the dummy's occupation and functioning in the same space as the viewer also assert its emotional gesture, its forlorn sense of being trapped in "the real" under the visitor's gaze. In such work, Oursler suggests, "[v]ideo no longer acts as a window to look through but is somehow made physical"[3] remarking later that, "I have tried to create entities [...] which have the properties of media space, put them in real space – see how they operate."[4]

These investigations of "media space" imply the performance of a kind of palimpsest in which real, virtual, and simulated spaces and events negotiate a writing over, reconfiguration, and translation of each other. As such, the key figure of such multi-media practice is *disjunction*: "a lack of correspondence or consistency" (*OED*) between terms that are nevertheless structurally bound one to another. When borrowed from architect Bernard Tschumi's definition of the intractable "internal contradiction" he sees as defining architecture's operation, this term extends the spatial metaphor by which multi-media performance and installation may be usefully read. Arguing that the experience of architecture is in its essence "about two mutually exclusive terms – space and its use or, in a more theoretical sense, the concept of space and the experience of space,"[5] he emphasizes that "architecture is made of two terms that are interdependent but mutually exclusive."[6] Transposed into the relationship between "live" and "mediated" acts and spaces, "disjunction" suggests restless and contradictory relationships between interdependent, but distinct orders of representation and experience.

Mary Gearhart's photograph of Willem Dafoe's performance of Theseus in the Wooster Group's *To You, The Birdie! (Phèdra)* (2002) alludes to such disjunctions and their effects (Figure 6). The performance explores an intertwining and overlaying of "real," "mediated," "recorded," and "represented" spaces and times. Invoking Racine's adherence to neo-classical unities of time and place, *To You, The Birdie! (Phèdra)* incorporates the "real time" of game playing, as the principal characters repeatedly return to a fast-paced game of badminton "live" on the stage, as well as video recordings and live mediations of performers. As a counterpoint to the different continuities implied by the play's dramatic unities and the game's "real time," the performance is continually fragmented and mediated, both in the soundtrack, which incorporates distortions and delays, and a system of mobile plasma screens that move slowly and smoothly in front of the actors to divide their bodies and performances. In the moment captured by Gearhart, Dafoe/Theseus's body is divided: split explicitly between screen and stage "presence." Indeed, this is a body whose division is multiplied; articulated not only through a division between "live" and "mediated," but also through the divisions of the mediated itself.[7] Hence Dafoe/Theseus is "seen" ("live" on stage) and, simultaneously, seen "being seen" ("live" on screen, *via* the camera). This mediation articulates Dafoe/Theseus's "being there" in a conjugation of different orders of representation: in the "theatrical" (represented) body of Theseus, now deferring to the "live mediated" body, which refers to the "real body" toward which the camera has directed its attention. Such a moment explicitly addresses the perception of the actor's "presence" through the media's unfixing of the performer's *place* and in the unsettled, disjunctive relationship between the spaces and actions in which it is composed. Indeed, in the disjunctions between live and mediated, real and represented, presence and absence, "Dafoe/Theseus" comes to occupy more than one position, while this mode of performance comes to be defined in the mobility and simultaneity of its signs.

Notes

1. Quoted in Jacqueline Humphries, "Jacqueline Humphries Interviews Tony Oursler on His Newest Piece: Seven Months of My Aesthetic Education Plus Some . . .," Tony Oursler Online, *see* http://tonyoursler.com/tonyourslerv2/main.html.
2. M. Kelley, "An Endless Script: A Conversation with Tony Oursler," in *Tony Oursler, Introjection: Mid-Career Survey 1976–1999*, ed. Deborah Rothschild (Williamstown, MA: Williams College Museum of Art, 1999), 53.
3. Quoted in Elizabeth Janus, "Talking Back: A Conversation with Elizabeth Janus Oursler," in *Tony Oursler, Introjection: Mid-Career Survey 1976-1999*, ed. Deborah Rothschild (Williamstown MA: Williams College Museum of Art, 1999), 72.
4. Quoted in Humphries, "Jacqueline Humphries Interviews Tony Oursler on His Newest Piece."
5. B. Tschumi, *Architecture and Disjunction* (London: MIT Press, 1994), 15–16.
6. Ibid., 48.
7. Samuel Weber, *Mass Mediauras: Form, Technics, Media*, ed. Alan Cholodenko (Stanford, CA: Stanford University Press, 1996), 116.

17 Embodiment

Peter Lichtenfels

"Embodiment" and "embodied knowledge" are terms that carry weight in the discourse of practice as research, because they signify the process rather than the analysis, the place of performance and making performance that research into practice seeks to discuss. The terms are widely used in dance and choreography studies, but far less so in the world of theatre. Matthew Goulish, Tim Etchells, and a growing number of others, have begun to describe with words on a page what I, as a theatre director, think of as words-and-body – an embodiment of verbal text similar to the embodiment of movement-as-text that is dance.[1]

To begin to talk about embodiment and theatre I have to turn to my practice: I am directing a production of Shakespeare's *A Midsummer Night's Dream* in Shanghai, in Mandarin. I know no Mandarin, the cast knows little English. I rely on the actors to tell me their understanding of the Mandarin translation; I share with them my understanding of the lines. We are rehearsing in the imperfect world of the translator and our translation.

I am alarmed, but have deliberately placed myself into this situation. I want to probe deeper into questions and concerns, one of which is this notion of *embodiment.* My understanding of embodiment is the practice of the body doing, and developing an understanding of that doing. As people, whether Chinese actors or occidental director, we understand ourselves through our body, whether within the limitations of gravity, or the form of our bone structure. *Embodied knowledge* is also the practice of the acculturated body within our social and cultural environment, but embodiment in my theatre practice means training another layer of body knowledge – which is why the theatre rehearses.

We begin working on the first scene between Hermia and Lysander (1.1.128–224) when one pair of lovers has just been told by Duke Theseus that they cannot choose each other as partners: first time through, the scene is terrible. Each body holds onto one tempo rhythm, emotes, and their spoken words are layered with a generalized tone. Lysander holds his body rigid, snatching at breath; Hermia is frantically acting "reacting."

The quality I value as a director, and as a member of the audience, is an energy greater than in daily life. Breath is the way words-and-body embody. And it is my contention that if the actors do not breathe on stage, the audience cannot breath. It all becomes stifled – and no one is breathing in this scene.

Over the next hours, and in repeated rehearsals of the scene, we work on understanding what the words might mean in our respective texts, Chinese and English, agreeing on a common meaning, problematizing it, probing what is being said but also what is not being said, coming to an understanding that both words and silence are powerful tools for persuasion, that the words have to be found. We work on the notion that because we breathe words out of the body, they can affect bodies physically, either in speaking or receiving them. This only communicates if we, the director/the audience, can see the body react. We agree on the need for instant transformation, a process in the moment, rather than demonstration of a process of transition.

Lysander and Hermia are good actors. Over the next weeks I constantly challenge, ask new questions, make them aware of physical habits. I am constantly amazed in the freedom, the presence of their bodies the moment we stop acting. I make them aware of the difference and insist on that presence for the scene.

My understanding of how I work with actors on embodiment is to visit, revisit, and keep revisiting all moments; constantly asking questions, always interrogating. The actors consistently create and absorb the knowledge articulated by their acting and they deepen the moments of words and movement. Through continual strategies of repetition the actors embody their knowledge.

We rehearse in this way so that they know there is so much work in creating a moment that each time they repeat the scene they will breathe and transform. If they do the work of words-and-body, each time will take more energy to hold breath than to breathe. Each time I rehearse the moment, the actors deepen their embodiment. We do not go to each scene to perfect repetition, but to prepare the body to "forget." As we keep working the scene the actors begin to realize that their understanding, word-speaking and movement, has become embodied, that they can trust this work; they can play and transform as the need arises, all in front of the audience.

We finish three weeks of rehearsals and Hermia has dropped her voice into her solar plexus. It has given her a freedom and concentration she has not felt before, and for the first time I want to listen. Because acting is so hard, for the last day or so she veers between the old and the new. But something profound is changing. Lysander has had a moment, maybe two, where I thought he would break through. But he is young and scared and self-protective.

I am preparing their bodies to *embody* the moment on stage, to breathe the words-and-body. And while I accept that I do so in a culturally specific way, I believe the embodied expressions of the actors' moments are loose enough to embrace their Chinese cultural traditions. Embodiment in theatre practice is a word signifying the way the actor trains to breathe the words-and-body so that the saying/thinking of the text happens for the first time for each audience.

Note

1. Matthew Goulish, *39 Microlectures: In Proximity of Performance* (New York and London: Routledge, 2000); Matthew Goulish and Stephen Bottoms, eds, *Small Acts of Repair: Performance, Ecology and Goat Island* (New York and London: Routledge, 2007); Tim Etchells, *Certain Fragments: Contemporary Performance and Forced Entertainment* (New York and London: Routledge, 1999).

18
Environment

Baz Kershaw

The incorrigible specificity of all theatre and performance rests in its inescapable interactions with an environment of some kind or another. This is why in the third Christian millennium the survival of all theatre, if not all performance, may depend on new ways of assessing the *environmental performance* of nations. Since 2005 the relative impact of 149 countries on the Earth's biosphere has been available through a UN-sponsored project – EPI: Environmental Performance Index – in a global ranking based on their environmental health and ecosystem vitality. Inevitably the EPI is contentious in its detail, but still it constitutes a profoundly significant factor for analysis of performance and theatre practice in the twenty-first century. For example, it provides a contextual tool for assessments of theatre/performance as part of the economic productivity of "cultural industry" activity in respect of a nation's GDP (Gross Domestic Product) or GVA (Gross Value Added) and therefore its potential contribution to climate change.

However, this approach to the environment–performance nexus may seem strangely contrasting to the standard accounts in reference texts for theatre and performance, which commonly assign the coining and popularization of the term "environmental theatre" to Richard Schechner in the 1960s/1970s. The focus of his seminal 1968 essay, "Six axioms for environmental theatre," overall is primarily on performers, spectators, and space – "all the space is used for performance; all the space is used for audience"[1] – which in his practice with the Performance Group led to a blurring of differences between theatre and ritual. Though the dynamic between scenographic and "found" space is part of this conception of environmental theatre, the emphasis is primarily on the *use* of space for experiment in human interaction, not on what the environment might *afford* to events created by humans and other organisms. This difference can be detected in the fact that the "found space" of the Performance Group's garage venue had a specially built set installed for each show they staged there. This human reconstruction of a "found" (or given) environment for human use is important, both for understanding the historical legacies of 1970s environmental theatre and for historiographies that would place it in longer disciplinary perspectives.

The twofold emphasis on spectator–performer and space–scenography suggests contrasting genealogies for this version of environmental theatre, the first routing back to widely spread religious, folk, carnival, and ritual performance traditions, the second focused through Western avant-garde experimentation back to eighteenth- and nineteenth-century European grand opera, to Elizabethan open-air theatres, medieval mystery cycles, Roman amphitheatres, and perhaps classical Greek theatre. The first indicates a diaspora of performance practices which relate very diversely – often through *interactive* responsiveness – to environments, as their traditions focus on the energies of human exchange in a vast range of locations that admit of deities, demons, mythologies, legendary figures, animal spirits via, for example, the enchantments of hieratic action. The second sketches a sequential pre- to postmodern *incorporative* scenographic/architectural "conquest of space" in which environments are shaped to human interests and desires, their materials often further subordinated imaginatively through cultural forms such as allegory, pastoral, landscape, wilderness, and "urban jungle." Of course such grandly dualistic lineages are bound to oversimplify the phenomenal plethora of how performance actually has been used by *Homo sapiens*. But, also, they can usefully suggest how the innovations of twentieth-century environmental theatres engaged with profound contradictions of human dwelling on Earth. The paradoxical human primate simultaneously performs both as master and servant, commandant and supplicant in an indispensable environment that is rendered increasingly unsustainable.

Have post-1968 practices added any positive perspectives beyond this performance–environment double bind? The genre most often genealogically linked to environmental theatre is site-specific theatre/performance, a term that gained international currency in the 1980s/1990s as increasingly experimental groups staged events which related to the specifics of particular places in spaces outside theatres. Often rooting for a sense of the local in reaction to globalization, site-specific events usually implicated the human figure in a situational ethics that paid heed to the history, archaeology, geology, or other aspects of a location. In tandem, the longer emergent coupling of "environmental" with ecological anxieties gained a growing performative profile through activism in protest events and pro-eco actions staged by burgeoning global social movements. But modernism's legacies were rabidly stubborn. Human pro-environment performance practices were often infused by cultural preconceptions linked to environmental exploitation: site-specific theatre identified space/place as "landscape" or "geography" or "archaeology" even whilst imaginatively searching out its natural affordances. Hence many innovative performance practices of the late twentieth century increasingly took direct inspiration from ecological history/philosophy as well as the new global environmental movements.

So, by the early twenty-first century there were growing signs internationally of a revaluation of "environment" in theatre and performance practices/studies from ecological perspectives and in light of the environmental imperatives of global warming. Hence performance and theatre at their best could become crucibles for clarifying antidotes to an endemic ecological calamity for humanity.[2]

Notes

1. Richard Schechner, "Six Axioms for Environmental Theatre," *TDR* 12, no. 3 (1968): 48.
2. *See* Paul Allain and Jen Harvie, *The Routledge Companion to Theatre and Performance*, Routledge Companions (New York and London: Routledge, 2006); "Ashden Directory," *see* www.ashdendirectory.org.uk/; Kevin Michael DeLuca, *Image Politics: The New Rhetoric of Environmental Activism* (New York: Guilford Press, 1999); "Environmental Performance Index," *see* http://epi.yale.edu/Home; Gabriella Giannachi and Nigel Stewart, eds, *Performing Nature: Explorations in Ecology and the Arts* (Bern: Peter Lang, 2005); Nick Kaye, *Site-Specific Art: Performance, Place and Documentation* (London and New York: Routledge, 2000); Dennis Kennedy, *The Oxford Encyclopedia of Theatre and Performance*, vol. 1, A–L (Oxford: Oxford University Press, 2003); Baz Kershaw, *Theatre Ecology: Environments and Performance Events* (Cambridge: Cambridge University Press, 2007); Patrice Pavis, *Dictionary of the Theatre: Terms, Concepts, and Analysis*, trans. Christine Schantz (Toronto: University of Toronto Press, 1998); Richard Schechner, *Environmental Theatre*, expanded edn (New York: Applause, 1994).

19
Lab/Studio

Shannon Rose Riley

Institutional cartographies – mapping the spaces of an interdisciplinary arts education in the United States

Undergraduate program in Fine Arts (at a BFA- and MFA-granting "art school"):

- Drawing studios
- Painting studios
- Sculpture studio
- Photography studio – darkroom (lab)
- Sound lab – recording studio
- Video lab
- 2D and 3D design rooms
- Wood- and metal-shops
- Ceramics workshop
- Performance space

MFA program in Studio Art (in the combined program of an "art school" and private university):

- Individual studio space
- Film studio – editing lab
- Sound lab – recording studio
- Video lab
- Media lab
- Metal- and wood-shops
- Performance space

Doctoral program in Performance Studies (in a department of theatre and dance at a large state university):

- Seminar room
- Research library

The spaces where PhD students were not supposed to spend too much time:[1]

- Acting labs
- Acting/devising/production workshops
- Dance studios
- The theatre

Labs and studios are distinct from academic research in the minds of many people, and have affiliations with a range of interdisciplinary meanings that obscure their political history. Unearthing these affiliations and meanings raises provoking questions about practice as research.

According to the *OED*, "studio" is related to the Latin verb *studere*, to apply oneself, to study. Commonly, it describes "[t]he work-room of a sculptor or painter; also, that of a photographer" – but there are also film studios, television studios, recording studios and studio apartments. The term "lab" is short for "laboratory" and is related to the Latin verb, *laborare*, to labor. It commonly refers to a space used for "conducting practical investigations in natural science, originally and especially in chemistry, and for the elaboration or manufacture of chemical, medicinal, and like products." A less common usage also found in the *OED*, unique in the early twentieth-century United States, is the "laboratory school" – a kind of "demonstration school," typically affiliated with a college or university and "used for educational research and in the training of student teachers and the demonstration of teaching methods." Studios appear to be spaces of personal study, of expression – labs, of labor and experimentation. Studios appear to be sites of creative production in the arts – labs, of scientific or technical experimentation and production. But this is oversimplified. In both modern visual arts and dance programs in the United States, studios are sites of experimentation, production, and the replication of findings and/or methods.[2] In theatre programs, labs are also sites of personal study and expression. Both definitions map spaces for study or investigation (research) and some kind of production (technical and/or aesthetic) as the result of labor.

In the popular imagination, "labs" are often pictured as "clean" and controlled sites, "studios" as uncontrolled, "dirty" spaces encrusted with paint and so on. Studios are imagined as sites of individual expression – labs, teamwork. The two appear to have different funding requirements: most scientific labs in educational contexts are funded by external grants and are imagined as more expensive than studios, particularly in theatre and performance programs, where studios are often considered little more than space by university administrators.[3] However, such stereotypes fall apart with interdisciplinary comparisons.[4] For example, sculpture or design studios are quite costly – they require specific materials and equipment, such as extensive ventilation systems. And despite notions of teamwork, the scientific lab has been the scene for displays of individual genius – and many visual artists and dancers working "in the studio" produce collaborative work. The lab is imagined as a site for reproducing replicable outcomes and the studio as a site for unique moments of expression – but this too is challenged through interdisciplinary comparison.

Educational institutions in the studio and performing arts in the United States are filled with "labs" and "studios." In undergraduate programs in visual or studio art, "studios," are commonly spaces where instruction and critique in drawing, painting, and sculpture occur; "labs" are sites of inquiry in electronics and "new" media.[5] Distinctions and overlappings between lab and studio activities and knowledges are telling. For example, in the early 1980s, the sound "lab" at a prominent art school contained wave-generating equipment; the recording "studio" contained multi-track recording decks, microphones, etc. for producing soundtracks. This complicates *OED* definitions – the sound lab is the site for experimentation with cutting-edge equipment, but the studio (rather than the lab) is the site for production, albeit one that may be aesthetic or expressive. Both require technical skill. Similarly, undergraduate photo courses often have two segments: studio and lab. The former consists of critique and lecture, the latter of time in the darkroom laboring to acquire technical skills and experimenting with results.

The fine arts studio recalls a time when learning art meant producing "studies." It might appear as the traditional site of art production (the expressive artist alone in his loft/studio). In contrast, the lab might be seen as the site of modernization, collaboration, and repeatable outcomes. However, even traditional studio activities partly meet this last criterion: producing "studies" requires an attempt to replicate and embody master techniques. Further, the modernization of schools of art and design in the United States is largely the result of Bauhaus influence and is at least partly concerned with the replicability of teaching art if not also with the replicability of determining the standards of "good" art and design.

The Bauhaus school, *Das Staatliches Bauhaus*, was formed by Walter Gropius in Germany (1919) as a place for experimentation and education in architecture, industrial art, and fine arts. Its aesthetic was fervently modernist. It united form and function and celebrated industrial fabrication and materials. Its pedagogical strategies were quite different from earlier approaches and were based on modern theories of design and specialization; traditional "studio" practices were modernized – all students took preliminary courses in design before moving on to specialist workshops. Many teachers at the Bauhaus relocated to the United States after the Nazis closed the school in 1933.[6]

Today, the Bauhaus method is the dominant mode of educating artists in the United States. Its impact is evident in required 2D and 3D design courses and upper-division studio work in specialized art areas/techniques. It lingers in the technical workshop spaces (metal, wood, and so on) that complete the lab/studio landscape. The drawing instructor only occasionally wears a white lab coat – but Albers color theory assignments and gray scales still line many art department hallways.

In undergraduate and graduate programs in theatre, however, the term "lab" often refers to a space used to explore the techniques and processes of acting. The emergence of the "acting lab" (rather than acting studio) recalls Grotowski's usage. In 1962, he added the term *laboratorium* to the name of his theatre company – and within four years, the group was recognized by the Polish state.[7] It has been suggested that the shift in terminology was partly to avoid the heavy censorship that professional theatres were subject to in Poland at the time.[8] Other benefits of this might include access to state funding resources. To be sure, turning the theatre into a lab makes tremendous sense in soviet Poland: Cold War logics privileged science and technology and Marxist theory takes science as a mark of rational modernism. As Ian Watson notes, Grotowski was tremendously interested in science, specifically the Bohr Institute.[9] Rethinking theatre as a laboratory can be seen as an attempt to theorize (perhaps legitimize – perhaps protect) the theatre arts from a modern, scientific perspective.

In arts education labs and studios overlap at times and repel at others. The only way to map the terrain is to consider what is at stake in such locations. How does the "studio" frame the artist as a lone genius and situate his or her space outside of society – not dissimilar from the ivory tower academic perhaps, both romanticized notions of genius?[10] How does the "lab" romanticize collaboration without troubling it? When might collaboration fold into corruption? Given the context of emerging fascism (Bauhaus) or Soviet communism (Grotowski), why is so little said of the modernization of the arts and German nationalism or the use of lab terminology as a way to negotiate power in the Soviet Union? Finally, why are labs and studios flexible in a way that doctoral program seminar rooms are not? What keeps us from speaking of the seminar studio, the discussion lab, the theory lab, or theory studio?

Notes

1. Soon after I graduated, the department began to offer both "academic" and "practice-based" doctorates. *See* Hunter in Part III on the UC Davis programs.
2. *See* Foster in this volume *re* the dance studio.
3. *See* Hunter in this volume.
4. *See* Sabatini in this collection on labs as theatres.
5. "Sound labs" typically include analog/digital equipment for sound generation; "video labs," equipment for image production and editing. "Media labs" increasingly consist of computers used for sound and video (both generation and production), various kinds of digital design, and interactive media performance, perhaps making individual sound and video labs obsolete.
6. For example, Mies van der Rohe and Moholy-Nagy resettled in Chicago; the latter founded the New Bauhaus school, which became the Institute of Design – part of the Illinois Institute of Technology. Albers taught at Black Mountain College and became head of the art department at Yale.

7. Leszek Kolankiewicz, "Grotowski and Flaszen–Why a Theatre Laboratory?," Nordisk Teaterlaboratorium/Odin Teatret (odin@odinteatret.dk), *see* http://www.odinteatret.dk/CTLS%20web/PDF/WHY%20Kolankiewicz-EN-pdf. pdf.
8. *See* Jeffrey Goldfarb, "Theater Behind the Iron Curtain," *Society* 14, no. 1 (1976); E. J. Czerwinski, *Contemporary Polish Theatre and Drama (1956–1984)* (Westport, CT: Greenwood Press, 1988).
9. *See* Watson in this volume.
10. For a critique of the studio artist, *see* Suzi Gablik, *The Reenchantment of Art* (New York: Thames & Hudson, 1995).

20
Medium

Susan Kelly

In the context of visual arts practice concerned with social or political issues in the United Kingdom, many artists have, for some time now, described their work as interdisciplinary, or have understood their practice as a particular kind of situated research. Yet, despite this and the dissolution of media defined departments within art schools, it is still almost impossible to answer the question, "what do you do?" without recourse to one, or a list of, media as a kind of anchor or legitimation device.

The term "medium" is embedded with many of the knots, problems, and possibilities for practice-based research. The term slides between being used as a synonym for "discipline," and being seen as a seemingly straightforward designation of material substance or modality used in a piece of work. The material, technique, or the mode of expression used by the artist is seen as something the artist communicates *through*. The medium operates as a channel, or as a *means* of manifesting thought, intention, or perhaps affect. This idea of medium as "raw material" or as a given mode of expression seems relatively clear and uncontentious. Yet, if we look at the syntax of action enabled by this use of the term, and the position of the artist it inscribes, we run into several problems.

Much practice-based research (hereafter, PbR) is concerned with questions of epistemology. Practice as a form of intellectual inquiry crucially allows us to consider different registers of social, bodily, material, unconscious, spatial, and visual experience as forms of knowledge. PbR is a powerful framework within which we can fundamentally question the ethics of knowledge, how we know what we know, from what position we know it, and produce situations and strategies for not just knowing something new, but more importantly knowing it in a different way. This proposition for what PbR might be necessarily implicates the subjectivity and position of the artist-researcher. Her own knowledge and conventional authority is questioned and at the same time the practice is opened up to work on new terrain. The "research turn" at its best opens up questions around ethics, knowledge, and method that allow artists to have meaningful collaborations with sociologists, anthropologists, documentary makers, historians, ethnographers, and researchers from other fields, rather than merely borrow their language and codes.

However, within such collaborations and encounters, the artist's authority and the stability of their position are often smuggled back in through an extended use of the term "medium." Think about the difference between the artist who "uses" photography and the photographer or the artist who "uses" performance, but wouldn't define herself as a performer. Such distinctions allow broad and flexible operating principles for artists and permit a move away from technical specialization and definition of practice according to media type. Yet, these distinctions can also, paradoxically ascribe a fresh and fortified authority to the artist. Consider a recent press release from the ICA in London: "*Double Agent* is a group exhibition featuring artists who use other people as a medium. The show contains work in a variety of media including video and live performance [. . .]."[1] While the press release shows its ambivalence by stopping short of listing "other people" as a medium in the list that follows the opening statement, it is an astonishing and yet not uncommon use of the term "medium" in the context of recent socially or politically engaged art practices in the UK. This conventional use of the term "medium" and the syntax it produces allows "other people" to become just another silent and passive raw material, another medium, like wood for example, to be exploited. Questions of ethics, the artist's position, and the ways in which she might engage with or produce knowledge of a social field in this instance become irrelevant in the quest to produce an ever-expanding list of media at the artist's free disposal.

One could say then that the term "medium" allows the syntax of conventional artistic practice – and its power relations – to remain undisturbed. In the case of socially or politically engaged practices, the artist's position remains authoritative and knowledge produced through the medium of "other people," in this example, is returned to the artist as a given. If the term, *medium*, becomes a blockage point, or a limit in the setting to work of PbR – a research modality that precisely questions the position of the artist and opens itself up to other kinds of knowledge and other ways of working – then are we left with no option but to abandon the term?

Perhaps, however, medium could also be seen in the scientific sense of a "pervading or enveloping substance; the substance in which an organism lives; *esp*. one in which micro-organisms, cells, etc., are cultured" (*OED*). "Medium" might become a term that is used to describe a substance that is both *formed* by the disparate knowledges produced in PbR, and a substance that *holds* such research in a place where it does not have to become just another object of knowledge or element in an artwork. The medium might become a generator and holder of a place in between, within which these sometimes tentative and fragile knowledges can emerge and develop a necessary consistency. A medium described in this way would have to be invented each time by a practice, proceeding internally, developing its own logic and texture. Such a medium could structure the activity of PbR without disciplining it through tying it back to safe havens of anchorage and legitimacy.

Note

1. "Double Agent." London, 2008. Institute of Contemporary Arts. 17 August 2008. *See* <http://www.ica.org.uk/Double%20Agent+15880.twl>.

21
Oral History

Della Pollock

Oral history began primarily as a method, propelled by the technological innovation and popularization of the tape recorder, which researchers began using in the 1930s to supplement archives with otherwise unrecorded historical materials. Initially, oral history focused on broad historical events, periods, and thematics; subjects tended to be elites with personal archives and rich memory banks that could be, and conventionally were, externally validated. By the late 1960s, scholars like Paul Thompson began to discern its radically democratizing potential, and the possibility not only for filling gaps in official records but also for expanding those records to include the stories of people and experiences buried under or excluded from histories of conquest.[1]

These histories had not been previously recorded in part because they belong to what Diana Taylor calls "a repertoire of embodied thought/memory."[2] For Taylor, the repertoire, as opposed to the archive, is a body of generally debased remains; it "holds the tales of the survivors, their gestures, the traumatic flashbacks, repeats, and hallucinations – in short, all those acts usually thought of as ephemeral and invalid forms of knowledge and evidence."[3] The "new" oral history consequently presented political and methodological problems for historians, requiring them, to some extent, to forego validation of memory-based claims by archival documents; to shift from a text- to a subject-centric history; and to favor perspective and perception.

Changing subjects changed the terms of historical understanding – from *validity* to *value*, from *demonstrative* (or evidentiary) to the kind of *narrative* truth Trinh Minh-ha claims for postcolonial histories, the truth "in the in-between of all regimes of truth."[4] Narrative and subject-centered history (often dismissed as "subjective") thus has an explicit politics often obscured by more objective or, indeed, objectifying means of doing history. It requires the researcher to investigate the performative nature of the subject, to ask: *Who speaks? What difference does speaking make? To what extent is the much-heralded agency of the speaker conditioned by prior speech acts? To whom does her story belong? Whose interests does it serve?* Going well beyond reductive divisions between the telling and the told, as well as method and findings, oral historians began to grapple with the power dynamics intrinsic to the ethno-historical "encounter," asking: *How does the performance of the past in the present shape and make not only history but historical subjectivity?*

The rise of oral history *vis-à-vis* history echoed the rise of ethnography *vis-à-vis* anthropology. While both may have begun as methodological substrata, each grew to challenge basic disciplinary assumptions. Oral history and ethnography cross in the emergent role of the oral historian as an ethnographic participant in culture construed as a matrix of micro- and macro-performances.[5] From a performance-centered perspective on culture, oral histories are *reflexive, rehearsed,* and *renewed* narrative performances whose meanings, as such, are *emergent* and *contingent.* Like the ethnographer, the oral historian is also a representational agent of the cultural performances in which he/she participates; ideally, he/she is a performative co-witness.[6]

As early as 1982, Samuel Schrager displaced the foundational appeal of the "individual" story authorized by reference to "experience," locating oral history instead in an ongoing network of dialogic relations. Schrager observes that stories he heard during early research "had grown out of a conversation that [. . .] continues wherever there is talk about the memorable past." He argues: "There is no choice about being inside or outside the dialogue. Everyone is part of it, invested with a participant's responsibility for sifting through what he or she has lived."[7] Oral history is relational and dialogic. It is defined by the mutual investment of tellers and listeners in discerning value in lives lived and establishing social and personal *ethos*. Hayden White defines narrative by the ethical thread the narrator draws through it;[8] Butler emphasizes the way in which life narrative defines an ethics of relational investment.[9] The oral historian enters this scenario as one more participant in an ongoing conversation, rather than primarily as a recorder or chronicler. Slim and Thomson describe oral history as an art of listening that requires historians to participate in a horizontal versus vertical economy of narrative exchange that specifically challenges top-down approaches to community development.[10] They attribute the importance of listening in part to Alessandro Portelli's sense that oral history is a living negotiation of memory and desire, repetition and imagination. For Portelli, "the importance of oral testimony may sometimes lie not in its adherence to facts but rather in its divergence from them, where imagination, symbolism, desire break in."[11] In listening *for* those moments at which "imagination, symbolism, desire break in," oral historians listen *forward.* They attend to the implications of telling the past in the present for the course and force of change.

A performance-centered approach to oral history understands any one account as a critical repetition among repetitions; its liminal truth – "in the in-between of all regimes of truth" – as at least complementary to "the hierarchical realm of facts" typically favored by the social sciences;[12] the teller as authorized by prior tellings; the interviewer as directly implicated in a narrative environment that long precedes and follows his/her conversations; and oral history as a *changing* (shifting and transformative) mode of making meaning. It gains an additional political valence in the performance of retelling, in the dialogic reiterations of the interviewer as a responsible witness. Oral history is itself a performance in the sense of a multiply constituted, culturally rehearsed, relationally charged, and radically contingent practice of responsible witnessing.[13]

In practice, this means that the interviewer-as-audience member reflects back the micro-politics and ethics enacted at the relational nadir of the interview process. His/her story is or should be a reflexive account of tactical and sensuous dynamics that embody history working itself out in narrative interaction, *on*, *through*, and *by* interview participants. Listening for (a) change changes the political and epistemological landscape of (oral) history. It marks the past as only knowable in the present;[14] history as neither outside dialogue nor any historical subject as "outside" memory; memory as an art of remembering drenched in desire and imagination, often drowned out by violence and shame,[15] and yet driven by ethics and necessity to find recourse for violence and trauma, and hence to perform recursively towards redress.

Notes

1. Paul Thompson, *The Voice of the Past: Oral History* (Oxford: Oxford University Press, [1978] 1988).
2. Diana Taylor, *The Archive and the Repertoire: Performing Cultural Memory in the Americas* (Durham, NC: Duke University Press, 2003), 192.
3. Ibid., 193.
4. T. Minh-Ha Trinh, *Woman, Native, Other: Writing Postcoloniality and Feminism* (Bloomington: Indiana University Press, 1989), 121.
5. *Re* performance ethnography, *see* Barbara Myerhoff, *Number Our Days* (New York: Simon & Schuster, 1978); Victor Turner, *From Ritual to Theatre* (New York: Performance Arts Journal Publications, 1982); ibid., *The Anthropology of Performance* (New York: Performing Arts Journal Publications, 1986); D. Soyini Madison, *Critical Ethnography: Method, Ethics, and Performance* (Thousand Oaks, CA: Sage Publications, 2005).
6. Dwight Conquergood, "Performance Studies: Interventions and Radical Research," *TDR: The Drama Review* 46, no. 2 (2002).
7. Samuel Schrager, "What Is Social in Oral History?" [1982], in *The Oral History Reader*, ed. R. Perks and Thomson A. (New York: Routledge, 1998), 76.
8. Hayden White, "The Value of Narrativity in the Representation of Reality," in *The Content of the Form: Narrative Discourse and Historical Representation* (Baltimore, MD: Johns Hopkins University Press, 1990).
9. Judith Butler, *Giving an Account of Oneself* (New York: Fordham University Press, 2005).
10. Hugo Slim and Paul Thompson, *Listening for a Change: Oral Testimony and Community Development* (Philadelphia, PA: New Society Publishers, 1995), 5.
11. Quoted in ibid., 156.
12. Trinh, *Woman, Native, Other*, 121.

13. Della Pollock, ed. *Remembering: Oral History Performance* (New York: Palgrave, 2005).
14. Jacquelyn Dowd Hall, "'You Must Remember This': Autobiography as Social Critique," *Journal of American History* 85, no. 2 (1998).
15. Marc Augé, *Oblivion*, trans. Marjolijn de Jager (Minneapolis: University of Minnesota Press, 2004); Harald Weinrich, *Lethe: The Art and Critique of Forgetting*, trans. Steven Rendall (Ithaca, NY: Cornell University Press, 2004).

22
Site-Particular

Ilya Noé

Aside from having become yet another fuzzy mainstream genre, another conspicuous tag used quite indiscriminately as an inherent signifier of vanguardism, the term *site-specific* has become a veritable battlefield, a war site where opposing positions dispute the nature of the site and the "right" relationship of art and artists to it.[1] As a result, a number of alternate vocables have emerged as an attempt to forge and define more complex ways of engaging with site,[2] as well as to reclaim practices that aim to be considered socially and politically progressive. While some site-specific artists and theorists continue to argue for rootedness and the physical anchoring of a work of art to a specific place, use, or reading in order to defy hegemonic forces, and others champion nomadism and refabricate and relocate previously installed or performed pieces for the exact same reasons, I propose a *particular* approach to site. Site particularity is not a fixed set of oppositional methods or even resistance to rules based on *a priori* assumptions. It is open to changing geographies and based on the understanding that all strategies are context-contingent and as prone to co-optation as any dominant theories or practices on the landscape – no matter how "avant-garde" and oppositional they set out to be.

Even though the move towards *site-specificity* (and live performance-based work) has often been traced as propelled both by the desire to withstand a system which turns art works into commodities, and by a reaction to modernist ideals of autonomy and universality, a lot of the work that is labeled *site-specific*, has clear universalizing tendencies. The term *site-specific* is used to describe work that is determined by and physically bound to its environmental context – be it a natural landscape or an ordinary everyday space – while *site* is merely defined as the actual material location. This modernist formulation[3] still assumes the neutrality of space experienced by a universal viewing subject and has been problematized by artists who have approached site as historically positioned and culturally determined space, rather than as a mere collection of physical attributes. But while the concept of site has been auspiciously expanded to include the social, the economic, and/or the political, it has ultimately, become quite unspecific.

In contrast, a *particular* approach attends to site not as something that pre-exists either formally, conceptually, or ideologically, but as something that is constructed performatively out of the exchanges between artist, environment, and audience. It is an ongoing series of interrelational and open-ended processes: always partial, always situated, multiply layered, often contradictory and messy, and produced by active agents negotiating between all kinds of positions and working through all kinds of relationships. Neither located in exclusively conceptual/ideological zones, nor tempted by geographical essences or totally unified social consolidations, a *site-particular* position is about becoming situated and valuing particularities rather than holding the other to some ideal standard and/or pretending that specific experiences are the same everywhere. The *site-particular* is not about essentializing site by either physically hinging the work to a specific place, use, reading, or by materially unhinging it and pretending that specific experiences are the same everywhere.

Rather than superimposing a ready-made grid upon events and situations, or operating in a relativist flat field where anything goes, *site-particularity* offers a long-term dynamic of to-and-fro within locational possibilities and limitations, made and remade. Nor is *site-particularity* a tool with which to "understand" site, but rather a conversation, a process of collaborative construction: a reciprocal and simultaneous way of shaping and being shaped, a continual relearning, rehearsing, and improvising of different ways to recognize and respond in the moment and with full awareness as new situations are encountered. The *site-particular* does not merely look toward relativist/postmodernist alternatives as a reaction to universalist/modernist ideals, but toward situated work that reassesses, troubles, and avoids romanticizing either settler or nomad trajectories.

Notes

1. Miwon Kwon, *One Place after Another: Site-Specific Art and Locational Identity* (Cambridge, MA: MIT Press, 2004).
2. Site-contingent, context-specific, site-determined, community-specific, site-oriented, and so on.
3. Richard Serra's work can be read in this manner.

23
Situated Knowledge

Lynette Hunter

The term "situated knowledge" has two quite distinct disciplinary connections. The first is with a long twentieth-century tradition in studies of knowledge that comes from the margins of society, from black feminist thought,[1] from non-mainstream science,[2] and from those outside accepted forms of social communication.[3] The second connection is with the study of learning that takes place in the process of engaged observation and practice, such as craftwork in silver, or children's acquisition of language,[4] or more recently, computing skills. This latter use is tied closely to studies in tacit knowledge.[5]

The two disciplinary fields have come together in studies of knowledge deriving from practice rather than written verbal proficiency in rational logic and analysis – including the practice of writing. These newly valued areas of knowledge include not only centuries-old training systems in dance, acting, and other performance media, but also, for example, indigenous people's traditional knowledge. In common with traditional knowledge, situated knowledge systems are not closed. They do not prescribe sets of rules or fixed delineations of content.[6] They are intensely practical rather than pragmatic, and are intended to respond in sophisticated ways to quite different contexts. Unlike scientific knowledge in which the effect of the observer is often a "problem" and many experiments are devised in order to minimize it, in situated knowledge the whole point is that the observer is engaged. It is only through their engagement that knowledge can be manifested,[7] and the observer is both the practitioner who makes things and the audience or respondent.

To give an example from everyday life: the way that women think has often been referred to as "intuitive"; however, over the last two decades this knowledge has been shown to be firmly rooted in observational experience and the pragmatics of daily testing.[8] The book *Women's Ways of Knowing* has been immensely influential in this field. Intuitive knowledge is not usually articulated as a set of ideas, but is continually put into practice through interaction with others.[9]

An example from performance practice comes from Inuit teaching structures, in which storytelling is the most important verbal form of conveying knowledge. In storytelling the performer does not record an immutable "fact," nor do they try to persuade about a specific situation, nor do they enact an "essence" of meaning. The story-teller performs a story that only becomes knowledge when the listener retells the story in their own way, in their own context, a contextual rehearsal. With a story, the responsibility is on the listeners to hear, interpret, understand, and put into practice in their own specific situations.[10] This kind of knowledge is profoundly embedded in performance traditions from around the world and is quite different from "catharsis" which is experienced in the moment of the performance. A related twentieth-century concept is Brecht's *verfremdungseffekt*, which describes an audience response that does not become emotionally identified with the plot of a play, but rehearses the story in their own contexts in order to generate a specific political action.

One of the most well documented case studies of situated knowledge comes from social studies in science and technology which have explored the impact of increasing numbers of women scientists on the disciplinary field of the biological sciences, demonstrating the gradual shift not only in the subject matter of experiments, but also in the practices undertaken in the laboratory.[11] In the arts, we could look, for example, at work on visual art as research defined as "not characterized by an objective, empirical approach [...] [striving] for generalization, repeatability and quantification" but "unique, particular, local knowledge."[12] Many other examples may be found in research on dance. For example, recent choreographic work by Hilary Bryan through Rudolph Laban's "Space Harmony" analysis of movement, engages with the Fibonacci series in mathematics, in harmonic architectural relationships in crystalline forms (such as the icosahedron), and the 12-tone scale in music, through the body movements of dancers.[13] Bryan is articulating the situated knowledge embedded in human movement that enacts basic patterns that generate the more well recognized knowledge in other areas such as mathematics.

This last example raises a conundrum for situated knowledge: if the knowledge in engaged observation and practice is articulated in more quantifiable ways, is it suddenly inaccessible to contextual rehearsal – does it become "fact"? I would argue that the arts are made up of practices in which contextual rehearsal is always available, if not necessary. In the arts, situated knowledge becomes a situated textuality, knowledge always in the making, focusing on process but situated whenever it engages an audience.

Notes

1. Patricia Hill Collins, *Black Feminist Thought: Knowledge, Consciousness, and the Politics of Empowerment* (London: Routledge, 1990).

2. Sandra G. Harding, *Whose Science? Whose Knowledge? Thinking from Women's Lives* (Milton Keynes: Open University Press, 1991).
3. Lorraine Code, *Rhetorical Spaces: Essays on Gendered Locations* (London: Routledge, 1995).
4. Jean Lave and Etienne Wenger, *Situated Learning: Legitimate Peripheral Participation*, Learning in Doing: Social, Cognitive & Computational Perspectives (Cambridge: Cambridge University Press, 1991).
5. A. Janik, "Tacit Knowledge, Working Life and Scientific 'Method,' " in *Knowledge, Skill, and Artificial Intelligence: Tacit Knowledge and New Technology*, ed. Bo Goranzon and Ingela Josefson (London: Springer Verlag, 1987).
6. Louise Profeit-Leblanc, "Four Faces of Story," *Canadian Journal of Environmental Education* 7, no. 2 (2002).
7. Lynette Hunter, *Critiques of Knowing: Situated Textualities in Science, Computing and the Arts* (London: Routledge, 1999).
8. Ibid., "Listening to Situated Textuality: Working on Differential Public Voices," *Feminist Theory* 2, no. 2 (2001).
9. Mary Field Belenky et al., *Women's Ways of Knowing: The Development of Self, Voice and Mind* (New York: Basic Books, 1986).
10. Lynette Hunter, "Equality as Difference: Storytelling in/of Nunavut," *International Journal of Canadian Studies* Fall (2005).
11. For example, Donna Haraway, "Situated Knowledges: The Science Question in Feminism and the Privilege of Partial Perspective," *Feminist Studies* 14, no. 3 (1988).
12. H. Slager, "Methodology," in *Artistic Research*, ed. Annette W. Balkema and Henk Slager (Amsterdam: Lier & Boog, 2004), 12–13.
13. Hilary Bryan, *Ego Alter*, dance choreographed for the production of *Share* at Mondavi Center for the Arts, Davis California and COUNTERpulse, San Francisco, 2005–7.

Part III

Mapping PAR in the United States – Communities, Classrooms, Stages, and Holodecks

24
When Is Art Research?

Shannon Jackson

The interrogative title of this chapter is probably resonant in specific ways for a very specific historical moment in the history of higher education. Though it does not include the more familiar simile-making conjunction "as," it is certainly linked to a wider network of simile-making habits. "Practice as research," "performance as research," "art as research" are all phrases that can be found in special journal issues, program descriptions, and on the websites of academic initiatives that seek to align *research* with another activity that has been heretofore understood as *not research*. In the process, the simile's structure reifies research as a condition to which the former term aspires. Meanwhile, the construction also obscures a longer intellectual history of creativity that would not have needed a simile for any of these terms to be aligned. Medieval histories of craft, Renaissance histories of art and science, studies of African healing performance, and many, many other sites would position artistic practice and research as synonymous.[1]

I have only myself to blame for such reifications and obfuscations, not only because I am using this title for my own chapter but because I also used it to structure a year-long, interdisciplinary faculty working group and speaker series at the University of California, Berkeley that tried in various ways to explore the apparently benign, often flat-footed, potentially enraging question: "When Is Art Research?" I defended the decision to work under this title in part because of my experience researching and writing an earlier genealogical study of performance in higher education. In that book, *Professing Performance*, I found myself confronting the fact that my chosen field of study had consistently endured an equivocal position within the history of the academy.[2] For that reason, I found it important to re-ask flat-footed and contradictory questions precisely because the university contexts I inhabited continued to do so. I also found it important in order to vitalize and complicate my own sense of what the answers could be, to ask questions of terms and associations that might have become normalized for me, and to examine the assumptions behind values that I might have stopped questioning. I came away from that project with many animating principles that have since guided my own teaching and institutional life. Because several of them would also inform future work on the art as research questions, I will present a few of them briefly here:

1. The new just might be old; your experiment might be someone else's tradition. Conducting a genealogy of university practice inevitably means encountering its propensity to reinvent the wheel. Because many varieties of contemporary "turns" might well invoke principles from histories forgotten or from other fields unknown to differently positioned scholars, we should all be quite careful before declaring any practices "new" or "innovative" but investigate alternate histories for the resources they offer.
2. The definitions of fields and practices are contingent upon – and hence change with – particular institutions. Performance and other arts find themselves differently located at different kinds of institutions, and these differences affect how that field is understood as both a mode of critical inquiry and a site for developing different skills. Such definitions reciprocally decide whether a field has its own department or is a subfield of another, whether it is placed within a particular school or occupies a school of its own, or whether its infrastructure contains only libraries and classrooms or also contains studios, stages, editorial suites, drafting boards, sprung floors, screens, and darkrooms.
3. Interdisciplinarity is not the same as interprofessionality. The desire to move across disciplinary divides or across divisions of theory and practice does not in itself solve the question of how such moves are judged and received or how their national and international impact can be gauged. The poet might read theory; the literature professor might stage a dance. But questions about innovation in art research also imply venues that arbitrate its significance and innovation, requiring us to reconcile processes of peer review, reception, and "citation" in different fields as well as prestige of venue for publication, presentation, and exhibition.
4. The ill-fitting field can offer new angles on the processes of institutional legitimation. To occupy performance – and the arts in general – at a research university is to be tempted by defensiveness. Rather than reimagining oneself completely able to address all institutional terms of legitimation, one's ill fit can be used to ask new questions about epistemology, research process, and institutional mission. For instance, I found myself asking how performance could help us reflect on the mental and manual division that underwrote the history of the twentieth-century university, or how the demands of its collaborative models might actually provide a path for a contemporary humanities movement that seeks more collaboration and public engagement. The ill fit of the arts might well be turned to opportunity in the perpetually redefined university context.

Running this working group and lecture series was an expansion in a variety of new ways, particularly because I sought to work across art forms and art histories. While that multi-referential term "Art" brings a number of questions with it, it was for me a more expansive way to organize a necessary institutional conversation at Berkeley. It had become clear to me in *Professing Performance* that my own field could not fully stand in for the variety of theoretical and occupational histories of different art forms.[3] With our faculty working group, we could allow participants from a range of art fields to question together the role of the arts at a research university as well as to come to terms with the differences amongst our skills, histories, and ideas of what "practice" or "research" might mean. The structure of the Mellon-funded group allowed seven faculty members to apply to be admitted and, in exchange for a course release, to participate in weekly sessions. At the same time, I organized a 13-event lecture series that was open to the campus and non-campus public. The series brought in artists and critics from a range of fields – photography, performance art, institutional critique, community art, choreography, poetry, fiction, film, architecture, and experimental music – to reflect upon how they would use their own work to answer, unsettle, or redefine the question of art as research. The series itself also represented another kind of collaboration outside of UC-Berkeley in building a Northern California coalition around the arts – including California College of the Arts, San Francisco Art Institute, SFMOMA, Counterpulse, Yerba Buena Center for the Arts, UC-Davis, and UC-Santa Cruz – so that we could share resources to create residencies, events, and lectures and sustained conversations between ourselves and a larger art world.

As a reader might expect, the weekly faculty group turned out to be a kind of godsend for many of us. It provided that space for faculty from different departments – Music, Art History, English, Theater, Dance, Architecture, Art Practice, and Film – to engage with each other deeply, regularly, and to act on opportunities for cross-arts collaboration. In a climate where professors are rarely allowed to be students again, it also provided a space for us to take some risks, to be amateurs in each other's fields, indeed, to realize that we actually did not know as much about each other's fields as we might have imagined. The process included moments of connection and moments of disorientation, and I think it is fair to say that both the opportunity to connect and the opportunity to be disoriented felt incredibly welcome, even decadent at times. That sense of disorientation occurred not only when we read critical essays and histories in each other's fields, but most interestingly when we decided to spend time in the studio spaces of each other's departments. The theatre professor's awkward holding of a musical instrument, the filmmaker's self-conscious movement in the dance studio, the choreographer's diligent work through the critical practice of the art scholar, all of these were moments when we came to terms with the habituated differences in our modes of attention and in our working lives. The cross-media nature of the arts means that there are bodily differences in how we occupy space as we do our jobs. Those differences affect the daily routes we travel on campus and those we unwittingly avoid. They affect how we feel comfortable standing or sitting, how we see the same object, and how we see similarity in apparently different ones. Through these pragmatic encounters with art making and with different skills in art research, we realized that we were doing research on the protean nature of the arts and on the arts in the academy. By making the commitment to attend intellectually and physically to domains of practice that we did not know, we constructed our own kind of sensitivity to a longer and difficult history of cross-arts collaboration in both the university and professional worlds. Because this working group interacted with another institutional center on our campus, the Arts Research Center, the group also became a place for us to ask some de-normalizing questions about what the mission of such a center should be. Our own risks and missteps during the year made it clear why the institutionalizing of the arts on campus can be so difficult, why cross-arts conversations often do not go as planned, and certainly why conversations between artists and scholars so rarely do. These cohabitations across arts departments (for "conversation" is not always the right word for the exchange) required a level of vulnerability, a willingness to be bad at something, to be awkward, unknowing, and to fall on your face. It required vulnerability to have one's professional habitus disoriented, and we left with a renewed commitment to creating more contexts that allow such decenterings.

At the same time, I also recognized that our experience was the particular experience of a particular institution, the experience of being Berkeley arts faculty in a research university known for its scientific mission and without the comforts and limitations of the "art school" as a unifying category. Most of the faculty assembled were housed inside a larger school of the Arts and Humanities, an institutional moniker that differs from the many other monikers in which art research might flourish (for example, the Art School, the School of the Arts, the School of the Arts and Sciences, the School of Theatre, the School of Fine Arts). Hence, in light of the highly contingent institutional location of UC-Berkeley, I found myself articulating some of the ways that collaboration with humanities does certain things for the arts. I would submit, for instance, that in the humanities there is a higher tolerance for work that exists in a critical mode or interrogative mode. Hence, if research in the humanities is so very often linked to "critique," that is, to the posing of a perpetual question, there is a way that this mode and the practice of art dovetail particularly well. At that same time, I also became interested in the fact that the question "When Is Art Research?" could be answered differently depending upon which discipline's notion of research one employed. While certain trends in the critical humanities might bring forward certain kinds of connections to the arts, I could not forget that there are sometimes as many points of connection between arts practitioners and the social sciences. With the social turn in contemporary art and community performance, the labor of many kinds of socially oriented art practices are more like participatory fieldwork than they are like the methodologies traditionally associated with the humanities. Meanwhile, for musicians or choreographers who work in the domain of "reconstruction," art practice was interestingly more akin to the archival and historiographical practices of the historian. Meanwhile, other dimensions of art-making, its infrastructure, its spatialization away from a classroom and into a studio, for instance, have more in common with the labor practices of engineering. The work of the artist is often more akin to the collectively spatialized, experience of laboratory work than it is to the seated, computered, paper-stacked, and highly individuated occupational practices of the humanities professor.

Lingering on different ideas of what the question might mean led me also to explore some of the ways that our faculty group and visiting lecturers began to answer the question, "When Is Art Research?" First, it was interesting to note the disparity in response to the viability of the question. For some, the answer to the question was obvious: never. And this "never" came out of the mouth of an artist as often as it came out of the mouth of chemistry professors, engineers, or cognitive psychologists who came to some of our gatherings. For other artists, however, the answer was so obviously "always" that they thought it absurd to pose the question at all. As choreographer Reggie Wilson said, he heard the words "art research" as one phrase: "When is [...] Art Research?" His answer: "Uh, at 3 o'clock?"

I found myself identifying different registers at which one could answer the question. First, there is the *Professional* register that says that "art" needs to be "research" when an artist is up for tenure or is asked to put together a curriculum that needs to be approved from on high. It is research when a particular type of Foundation is looking or a certain kind of peer review is employed. Or, as colleagues in electronic music and composition, Ed Campion and Brian Kane suggested, "Art is research when you need grant money." Second, there is also what might be called the *Process* answer, which says that "research" is the necessary thing that precedes art making. In this frame, research is in the interviews and archives required of the community artist or in the music archives that precedes the act of musical reconstruction. It is the work with the genomist that precedes bio-art, or the work on the site that precedes the design of the building. In this frame, the simile structure of Art as Research holds because the artistic processes behind art mimic the research processes behind scholarship.

Finally, however, there is also what might be called the *Conceptualist* answer to the question, one that would see the art object or event as itself a form of research. And it was here that this apparently obvious, flat-footed question forced us to articulate the thing that was right under our nose, namely that so very much of what might be called "art since the sixties" has positioned the art object as a medium of critical exchange. Conceptualist work undoes the bounded integrity of the art object in order to produce moments of reflective interaction in the encounter itself. Indeed, a reminder of what Immanuel Kant and Michel Foucault after him said of the goals of critique helped us to see the connection between contemporary art and critique more clearly. For Kant, critique marked an epistemological shift away from attempts to know more or to know better and instead to ask how we know what we know, or, in his words, "up to what point can we know what we know?"[4] For Foucault, that reflection took on a political cast, asking us to take a critical stance on apparently normalized forms of power and regulation; critique was the "art of not being governed quite so much" or at least, the art of "not being governed quite like that."[5] Throughout our speaker series and working group meetings, we experienced artists and art-produced moments that variously engaged in this critical turn, whether they used art practices to question the integrity of art forms themselves or whether they used art practices to question the integrity of certain political forms as well. Contemporary art, like critique, is often about asking us to question how we know ourselves to be where we are. The paradox on the administrative level of the university, however, can be that the spectator does not always embrace the conceptualist turn in the arts. And, when it comes to the evaluation of the arts at a science-based university, it might be precisely a conceptualist "research turn in the arts" that makes art look less like art to the chemistry professor or the senior administrator who wants to know why artists are on her campus in the first place.

Hence, when I was appointed to chair another ad-hoc committee composed by the dean to give guidelines to our tenure evaluation committee about "how to make sense of these artists," I found myself trying to answer this question "When Is Art Research?" from an administrative consciousness as well. I found myself trying to formalize parameters for taking in the deformalization of the autonomous art object, for allowing non-arts faculty to accept and celebrate the possibility that art can be doing research precisely because it violates a predetermined sense of its forms and its boundaries. Working with a slightly larger team of faculty representing different arts fields, we found ourselves articulating non-academic, but still traditional principles for gauging the prestige and impact of an artistic body of work – profile of venue, prestige of commissioning agency, prestige of gallery representation, competitiveness of awards, honors, and touring sites, the difference between a group show and a solo show or between a retrospective and a reconstruction or between a festival and biennial. At the same time, we needed to qualify traditional indicators of prestige by allowing for certain kinds of innovation in community art practices or vernacular forms expressly sited away from high-art venues. At times, we analogized artworks to scholarship, calling the former an alternate form of publication. At other times, such analogies needed to be resisted, especially in the case of certain kinds of performing art forms that – because they required the dancing or acting or cello-playing presence of the artist – could not be printed and mailed for so-called "publication." It was in those times, too, that we chose to foreground the dilemmas of durational and embodied forms that made it harder to reconcile the co-present demands of teaching with the co-present demands of art making. Finally, we could analogize the single artist to the single scholarly author only to limited degrees, asking for an awareness of the fact that new collaborative models of art-making (for example, social practice) and old collaborative models of art making (for example, theatre) necessarily flouted research-based assumptions of singular or even principal authorship. Finally, with these kinds of answers to the question of "When Is Art Research?" I found myself riding several more paradoxes around the issue, writing a document that will be used as a tool in faculty evaluation and faculty governance, while simultaneously trying to suggest that we need not be "governed quite so much" or at least, "not quite like that."

Notes

1. Katy MacLeod and Lin Holdridge, eds, *Thinking through Art: Reflections on Art as Research* (London: Routledge, 2006).
2. Shannon Jackson, *Professing Performance: Theatre in the Academy from Philology to Performativity* (Cambridge: Cambridge University Press, 2004).
3. *See* Howard Singerman, *Art Subjects: Making Artists in the American University* (Berkeley: University of California Press, 1999); Sullivan, *Art Practice as Research: Inquiry in the Visual Arts* (Thousand Oaks, CA: Sage, 2005).
4. Immanuel Kant, *Critique of Judgement [1790]*, trans. James Creed Meredith (Oxford: Oxford University Press, 2007), 71.
5. Michel Foucault, "What Is Critique?," in *The Politics of Truth*, ed. Sylvère Lotringer and Lysa Hochroth (New York: Semiotext(e), 1997), 29.

25
The Oral History Project: Practice-Based Research in Theatre and Performance[1]

Lara D. Nielsen

A spectre is haunting the halls of the academy: the spectre of oral history.

Alessandro Portelli[2]

I like it that Portelli invokes Marx for the thinking of Oral History methodologies (borrowing that iconic sentence, "a spectre is haunting Europe, the spectre of communism," from the first lines of the 1848 *Manifesto*) because this phrasing makes explicit for oral historians and arts researchers alike that central task of thinking about *the relations of production*. Oral history differs from other historiographic modes in that it privileges the shared narrative authorities of speaking subjects in relationship to one another – valuing the vernacular propinquities of highly particular and historically located voices which may otherwise remain marginalized by the hierarchical conventions of power that organize social activities.[3] This includes teaching and research activities that usually narrate and authorize the fields of theatre and performance.

At a glance, primary sources about artistic (often collaborative) practices in the performing arts deploy a variety of distinctive modes to put down on paper what it is that constitutes the "doing" of theatre and performance research. Looking beyond the current era of the intense professionalization of research in the arts, where "the peer-review article" is the dominant formal protocol for communicating and documenting research, researchers find that key published works about artistic practice present ways of translating essentially *oral practices* into various kinds of documentations and narrations, which are not always formally recognized as "research" manuscripts. From Stanislavsky's novelistic storytelling and journal writings, to Meyerhold's diagrams, Grotowski's "laboratory" reports, Atlantic Theatre's "practical aesthetics," and Mary Overlie's "Viewpoints," students of theatre and performance encounter a wide array of primary sources. Students and faculty both know, by contrast, that research institutions rarely permit such creative forms in their own work.[4] This is not an argument towards anti-intellectualism, a sometimes uncomely feature of the artistic research environment. Rather, the point here is to acknowledge *the challenges of narrating the shared and collaborative authorities that define the relations of production in artistic research practices*.

The good news is that despite all the accomplishments of theatre and performance research, we do not know, definitively, how to represent and theorize practice-based research (PbR): the question of narrative technique remains open. Using oral history, researchers have at their disposal a technique for engaging and documenting the nuanced relations of production that go into the exercise and exploration of their arts. From the perspective of the academic "outsider" in the humanities, oral history has long been recognized and deployed as an especially apt form for documenting and researching the performing arts; indeed, oral interviews are often the principal means of accessing information about and interpreting artistic research. But what if the oral historian is also a participant in the artistic field? From the perspective of the practicing "insider," talking with researchers is to talk among ourselves *as researchers* about what we do and how we do it. An oral history approach offers techniques for narrating and curating creative processes that may better reflect the work of PbR.

Using oral history in this way contributes to the objectives and values of PbR in the arts by creating a communicative space, a methodology, and the creative leeway in which it becomes possible to apprehend and value what can be understood as the field of practice itself, in a systematic way. Given the dystopic old norms of page *versus* stage relations of production – one of the dualistic aphorisms yet hobbling the field of theatre across US campuses – this is no small thing. Within the many divisions of performance research, the steady infiltration of exciting key terms (including *practice-based research/performance as research/applied research/ devised research*) suggests that this is a period of lively "investigation into feasible working methodologies" in the performing arts,[5] and across a variety of disciplinary and institutional sites. If artistic practice is finally beginning to be recognized as a legitimate form of research within the context of contemporary university and college spaces – a big if, as any survey on the allocation of budgetary resources will tell – then among the challenges of PbR is to find ways to substantially value, in educational institutions, the creative processes that may (and may not) privilege the goals of public performance.

Curricular inaugurations

In 2005 I began teaching *The Oral History Project* (*OHP*) in the Theatre Studies curriculum within the Department of Drama at NYU's Tisch School of the Arts (TSOA), in an effort to supply the time and methodological training to attend to such collaborative processes of artistic study, self-reflection, and creative documentation.[6] The NYU curricular mission promises to join professional training for actors, directors, singers, dancers, and technicians with professional training in the multidisciplinary history and theory of theatre and performance studies. In classic if correspondingly problematic form, the department is something like a United Nations Assembly, representing different methods and ideas about work in theatre and performance, in order to nourish and develop the work of its many constituents. All of this occurs within the context of overwhelmingly North American institutional traditions, organizing academic and conservatory practices alike. Over the course of the BFA program, students navigate this plenty with the hope of finding out who they are going to become in relationship to the fields of theatre and performance. For faculty and students, the opportunity is to find ways to work against simplistic models of professionalization that box work in theatre into the domains of practical and academic training, and to pursue developmental questions about the places and methods of *intellectual* and *artistic* work that hold tantalizingly between. The affirmation of professional training in academic and studio "communities" invites (and cannot guarantee) the opportunity for innovative pedagogical approaches to the theatre studies curriculum, against Cartesian divisions of labor between the work of the mind and of the body.

Deploying oral history methods for rethinking the purview of theatre studies aims to reassess what kinds of research environments might be productive to illuminate *the conversations, practices, and processes of theatre and performance research* for the task of historical interpretation, if by "history" we include the history of the present. In *OHP*, students learn to locate and address their PbR activities by studying their practices in conjunction with their interviewing experiences; each addresses the narrative contexts of their own artistic and intellectual training processes. They also learn to speak for themselves, as authoritative participants in such critical and creative processes. Developing the arena of PbR is akin to recent discussions of pedagogical practices in the arena of dramaturgy. For example, Carey Mazer suggests the goal of dramaturgical pedagogy is to "acculturate the students to a new role" and provide new "materials, insights, resources, documents, perspectives";[7] Judith Rudakoff especially raises questions that are relevant to oral history work: "We do not simply ask 'Who am I?' in this process. We ask, 'who am I and where is here?' "[8] The value of examining their own work and training as collaborative enterprises reveals students as participants in (and not simply receivers of) the field. Using a central premise of oral history, to approach PbR as constituting an ecosystem of living histories is to recognize and empower the role of students in their sites of residence: as critical producers of the institutional ecosystem, and "active cultural media" of importance for PbR.[9]

It further strikes me that BFA training in theatre and performance could be thought of as a sequence of field researches in the rigors of artistic and academic traditions. To this assessment the question remains: how do we equip students to engage with this participant field/work? The territorialities and distinctive properties of engaging PbR in this way evoke the ethnographer's eye: each school of training, each performance *practice* has its cultures, vocabularies, values, and constituencies. Without being alerted to it with such language, theatre and performance students everywhere engage in a series of field assignments, studying, observing, and participating in the cultures of artistic and academic training. Looks like field/work, sounds like field/work: *participatory* field/work. Located in the academy, in studio, in rehearsals, or in performance, theatre and performance practitioners move in and out of each other's working spaces, curating, crafting, reflecting, rehearsing, teaching. Field/work – or PbR – emerges in theatre and performance research as a matrix of intersubjectivity and narrative contestation, in which pursuit of the control of meaning remains a critically creative endeavor. James Clifford's examinations of ethnographic writings share with oral history the attempt to nuance the narrative authority of the subject's voice by prioritizing speakers as situated narrators in their own rights.[10]

The uses of interviews in PbR to develop faculty and student understanding of their work in the field opens the interdisciplinary training processes. Used in this way, oral history techniques highlight the careful examination of dialogues and participant observation within the broader narrative design of the student "as historian." The sophistication and low-cost availability of sound-recording technologies, moreover, facilitates an appetite for research detailing the transmissions of oral practices as both art and artifact. In short, the labors of talking, interviewing, and writing are time-consuming, ethnographically marked, and highly interpretational; they require thinking about *writings*, as well as recordings, as polyphonic narrative endeavors that may contextualize, illuminate, and interpret conversations in the arts. Whether separately or together, these documents may be applied as primary source materials, for reflection and analysis. Linear order, in particular, collapses in the contestations of intersubjectivity, so that the time of knowing – of speaking, listening, writing, interpreting as unstable referents – might delink guaranteed narratives about the past, in favor of its futures. "What would it require," Clifford asks, "consistently to associate the inventive, resilient, enormously varied [societies] with the cultural *future* of the planet? How might ethnographies be differently conceived if this standpoint could be seriously adopted? Pastoral allegories of cultural loss and textual rescue would, in any event, have to be transformed."[11]

In the context of theatre and performance research, Clifford's remarks appreciate with particular importance. Narratives about New York theatre and performance genealogies drive quintessential institutional identifications, whether concerning remembrances of the impacts of the epochal arrivals of Stanislavsky, Strasberg, the Group Theatre, Stella Adler; or the establishment of the TSOA in 1971 and the legacy of the Experimental Theatre Wing. Accomplished speakers, artists and critics alike, regularly hail the halcyon 1960s and 1970s in narratives of social and aesthetic radicalisms that simultaneously inspire and depress student audiences, nearly 50 years later.

There are a number of ways that faculty and students can help each other breach the gap between what it might have meant to be a student of theatre and performance "then" and what it might mean to research theatre and performance "now." A persistent nostalgic gaze risks doubling as an ethnographic narrative of the past that Clifford critiques for its asphyxiating foreclosure of futurity, placing contemporary researchers in the awkward anachronisms of a keenly felt bind in which the machineries of capitalist market forces sometimes figure as the only language of commonality. Placed against the conventions of "academic" training, faculty and students may find themselves stuck between the proverbial rock and a hard place. I do not think we have run out of ideas about PbR/PAR (this volume attests to the contrary), but in academic settings, the framework for PbR demands our most rigorous interventions.

Strategies

What was once the Achilles heel of "the bastard art"[12] – its inter- and intradisciplinary unruliness too often generating condescension from related fields in the humanities – now seems poised to become one of its key institutional strengths. Research in the performing arts has long modeled the features of what are now hailed as innovative frameworks for pursuing community engaged scholarship and teaching: as a field concerned with practice-based activities, with the force of aesthetic and intellectual histories alike, *and hence with the communities that sustain them*. Scholars, artists, and teachers regularly tread into each other's fields, engage in collaborative research, work with people outside the educational institution, and interpret the dynamics of such work for their own distinctive purposes. What becomes exciting and new for emerging interdisciplinary liberal arts pedagogies is, instead, tradition in the arts. Glancing at the broader view of contemporary institutional research and pedagogical concerns: the Association of American Colleges and Universities and the Carnegie Foundation for the Advancement of Teaching Statement on Integrative Learning, for instance, each support the possibility for applied research to change "the working relationships of the disciplines within an institution."[13] What were once called "service learning" models lean now towards collaborative public research activities in a wide array of fields. In the call for a "new revolutionary educational philosophy,"[14] and in the re-evaluation of what Ernest Boyer calls *the priorities of the professoriate*,[15] we can recognize an old friend in theatre and performance research: the praxis of PbR.

Decades past, applied and community-based theatre research has had to advocate and elaborate its cause: making the argument that art not only does, but must, happen in collaboration with public communities.[16] Della Pollock beautifully discusses the uses of oral history for making performances in the tradition of community-based and applied theatre.[17] Disciplinarily, oral history methodologies seem to "make sense" for community-based work. But I am suggesting that we not only engage oral history methodologies for PbR in all kinds of theatre and performance areas, but that we do so with and through the support of institutional configurations that increasingly seek ways to prioritize PbR in its many disciplinary manifestations. Macalester College, where I now teach, lists *OHP* within its Civic Engagement rubrics, allowing PbR/PAR to become more recognizable to portions of the institution that are not often asked to imagine theatre and performance as part and partners of a broader interdisciplinary and practice-based research mission. It is clear, moreover, that artistic communities figure as distinctively productive constituencies and publics in their own rights. Current research and pedagogy paradigms are finding ways to attend to such collaborative and creative promise in educational contexts that in turn effect the future of creative work in the fields of theatre and performance.

Nonetheless, corporatizing university efforts to distinguish between research and teaching activities exacerbate "theory/practice" divisions of labor that do not serve work in the performing arts well. At NYU, for example, tenure-stream faculty are considered "researchers" and everyone else as "teachers."[18] Despite such hierarchizing propaganda, and insofar as it remains the purview of the faculty, curriculum development has always been one of the available faculty mechanisms to articulate desired shifts in the organization at the local level. Curriculum is not foreign to research, but rather elaborates theoretical agendas. Faculty regularly enact methods of knowledge transmission and critique alike, passing on disciplinary traditions and an awareness of their constructions "before the law" of the field. Oral history practices supply one exciting possibility for pursuing a collaborative praxis – and a politics of friendship – in PbR, helping faculty and students to address the rigors of theatre and performance.

Notes

1. A version of this chapter was shared with the Performance as Research Seminar at the 2006 American Society of Theatre Research (ASTR) Conference, Chicago. Many thanks to Kris Salata and Lisa Wolford Wylam for organizing the seminar, and to its participants for their comments.
2. Alessandro Portelli, "What Makes Oral History Different," in *The Oral History Reader*, 2nd edn, Robert Perks and Alistair Thomson (New York: Routledge, 2006), 33.
3. *Re* oral history practices, *see* the collection mentioned in the note above as well as the Oral History Association website: www.dickinson.edu/oha/.
4. For a history of the field of performance in the USA, *see* Shannon Jackson, *Professing Performance: Theatre in the Academy from Philology to Performativity* (Cambridge: Cambridge University Press, 2004).
5. Bruce Barton, "Points of Departure/Point of Departures: The Perspectival Shift of Practice-Based Research," presented at the American Society of Theatre Research Conference (Chicago 2006), 11.

6. Particular thanks to Robert Vorlicky for supporting this project and to my former colleagues at NYU for their feedback.
7. Cary M. Mazer, "Dramaturgy in the Classroom: Teaching Undergraduate Students Not to Be Students," *Theatre Topics* 13, no. 1 (2003): 136–7.
8. Judith Rudakoff, "The Four Elements: New Models for a Subversive Dramaturgy," *Theatre Topics* 13, no. 1 (2003): 150.
9. My students listen to audio recordings of oral histories and read the written transcripts. They consider what is learned from the sound of the speech in the conversation and from the transcripts of the conversation in order to learn to listen for the differences between sonic and written records.
10. James Clifford, "Fieldwork, Reciprocity and the Making of Ethnographic Texts," *Man* 15, no. 3 (1980).
11. James Clifford and George E. Marcus, eds, *Writing Culture: The Poetics and Politics of Ethnography* (Berkeley: University of California Press, 1986), 115.
12. Susan Harris Smith, *American Drama: The Bastard Art* (Cambridge: Cambridge University Press, 1997).
13. Judith A. Ramalay, "Community-Engaged Scholarship in Higher Education: Have We Reached a Tipping Point?," in *Community-Engaged Scholarship for Health Collaborative Invitational Symposium* (2007), 3.
14. Ari Sitas, "Inqola Masondosondo! For a New Sociology of Civic Virtue," *Journal of World-Systems Research* 1, no. 2 (2000): 885–6.
15. Ernest L. Boyer, *Scholarship Reconsidered: Priorities of the Professoriate* (New York: The Carnegie Foundation for the Advancement of Teaching, 1990).
16. In the USA, *see* the work of Jan Cohen Cruz and Sonja Arsham Kuftinec.
17. "Oral history performance is strung between reference to real events and real listener/witnesses, between recollection and anticipation of historical change," D. Pollock, ed., *Remembering: Oral History Performance* (New York: Palgrave Macmillan, 2005), 7.
18. John Sexton, "The Common Enterprise University and the Teaching Mission," New York University, *see* http://www.nyu.edu/about/sexton-teachingmission04.html. Unlike other universities, arts faculty at TSOA are not tenure-stream.

26
University Gamelan Ensembles as Research

Henry Spiller

Music is the universal language of mankind.
Henry Wadsworth Longfellow[1]

Universal language – not!

Music is not the "universal language" touted by the likes of Longfellow. It is true that sonic artifacts many categorize as "music" crop up the world over, and outsiders often find an Other's "music" to be not only decipherable, but even sublime on their own musical terms. However, the kinds of meanings that sonic elements, musical gestures, and sound-producing activities produce are not consistent from community to community; outsiders are likely to misattribute meanings to the sound artifacts of Others that were not intended and would not be understood by an insider. One might as well say, "Language is a universal language." Upon hearing a strange language, people generally recognize that something is being communicated, but are not fooled into imagining they can understand its meaning. Only with translation can the meanings of other languages be conveyed. Music is different; an Other's music is more likely to be blithely *mis*understood than to be regarded as unintelligible gibberish, precisely because it gives the false impression of familiarity.

As an ethnomusicologist, I conduct ethnographic research that explores differences in human sound-producing behavior and sonic artifacts to contribute to a more nuanced understanding of the place of such activities in human cultures. Ethnography is, in essence, a kind of translation; a successful ethnography renders one group's cultural practices more intelligible to another. The overriding goal of all my scholarly acts – research, writing, and teaching – is to foster in my interlocutors a wonder for the variety of human behaviors and a healthy skepticism of their own gut reactions to them. My own work focuses on Sundanese music and dance from West Java, Indonesia; my research methods have been mostly qualitative, and the results of my investigations have been disseminated in conventional scholarly ways – primarily in articles, books, and conference papers.

Like many ethnomusicologists of my generation, much of my understanding of the music I write about is informed by performing. Since the 1960s, many ethnomusicologists have advocated performance as research – more specifically, the practical study of unfamiliar musical performance practices as a means to gather ethnographic data. One side effect of this sort of performance as research has been a proliferation of "world music" ensembles in Western colleges and universities, in which interested general students attempt to learn the nuts and bolts of unfamiliar musical traditions and typically mount performances for the local community. *Gamelan* – bronze percussion ensembles from the Indonesian islands of Java and Bali – have proved to be particularly popular in American higher education and have burgeoned in the United States over the past few decades.

My academic duties have often included teaching gamelan performance to college students. For some time I regarded these teaching duties as a mere epiphenomenon of my ethnographic work; after all, since these ensembles involve only outsiders to the musical traditions (me and the students), how could they possibly generate significant ethnographic data? More recently, however, I have come to regard these student ensembles as both sites for and products of my scholarly inquiries. This chapter will give an account of some of the ways that the student gamelan ensembles I direct have become more than a serendipitous by-product of research – they have become scholarly acts in their own right.

Beyond bimusicality

The influential ethnomusicologist Mantle Hood, among others, promoted "bimusicality" as a research methodology; he argued that learning to perform outsiders' music at some level is a vital component of ethnomusicological training, necessary in order to: (1) develop an ability to perceive nuances not typically developed in western musical training but potentially important in other musical aesthetics;[2] and (2) to facilitate communicating with potential insider consultants using the mode of "musical discourse."[3] Ideally, an ethnomusicologist might develop bimusicality during an extended period of fieldwork. Hood also pioneered the incorporation of practical musical training into preparation for fieldwork at home; during the 1960s, his Institute of Ethnomusicology at UCLA was the site of a number of "study groups" – ensembles, often led by an "imported" master musician – in which ethnomusicology students cultivated skills in musical discourse.[4]

My introduction to gamelan in the 1970s was exemplary of the sort of "bimusical" experience that Hood imagined. As an undergraduate student at the University of California, Santa Cruz, I enrolled in the music department's gamelan class, which was taught by an "imported" Indonesian musician, Undang Sumarna (as of this writing, Undang continues to teach gamelan at UCSC). Undang's approach was to name the piece, and then show each student, in turn, his or her part. Many gamelan pieces consist of a short melodic motif that is repeated over and over; each of the different gamelan instruments plays a variation of the motif that conforms to the particular instrument's idiom, so once a student learns a variation, it can be repeated *ad libitum*. Once a student appeared to have mastered the instrument's part, Undang proceeded to the next student and taught the next variation; if somebody slipped out of sync – a frequent occurrence – Undang returned to the offending student and corrected the part. Undang constantly darted from one student to another in an attempt to get all of them to play correctly at the same time.

Playing gamelan did not strike me as very difficult. It was more like a game than music making: specifically, a game called *Simon* – a 1978 electronic version of the children's game "Simon says." It played a random sequence of pitches using four colored buttons that lit up as they sounded and the object was to precisely repeat the sequence by pushing the buttons. Undang showed me 16- or 32-note sequences, which I memorized quite easily. Or so I thought; Undang returned to me with startling frequency. Sometimes my error was simply falling out of sync with the other musicians because I could not hear how the parts fit together. Sometimes, the error was conceptual; the patterns I discerned in the sequences of notes often were just plain wrong. An example: one of the first pieces Undang teaches his new students is called *Bendrong*. Undang generally used the digits one through five to name the pitches of the *salendro* scale; here's what he showed me for the *saron* (6-key metallophone):

111311131454322522252225245431131113111314543225222522225245 431131[...]

I heard a metrical accent every fourth note in the sequence, starting at the beginning, so my first attempt at patterning resulted in something like: 1113/1113 [...]. When I noticed that a similar pattern emphasizing pitch "2" emerged (2225/2225), I felt confident that I had cracked the code. I also noticed that there was an anomalous motif between the "1" patterns and the "2" patterns. So I repeated something like the following (I have inserted spaces to demarcate the patterns I heard):

three "1" patterns:	1113 1113 1113
anomalous pattern:	1454
three "2" patterns:	2225 2225 2225
anomalous pattern:	2454
three "1" patterns:	1113 1113 1113

anomalous pattern: 1454
three "2" patterns: 2225 2225 2225
[. . .]

Careful readers will have already found my mistakes; in my zeal to create three consistent "1" and "2" patterns, I ignored the pitch "3" that characterizes the anomalous pattern (substituting for it in some cases a pitch "2" or pitch "1"). Furthermore, by repeating my "1" pattern three times from the outset, I lengthened the opening phrase and immediately got out of sync with the ensemble.

I was on the right track with the notion of three "1" patterns and three "2" patterns, but needed to adjust the way I thought about rhythmic accents and their relationship to the patterns. In contrast to much Western music, beats and phrases in gamelan music are "end-weighted" – the stress or accent comes at the end of a unit, rather than the beginning, and any subdivisions of a beat precede (rather than follow) the beat itself. Regrouping the sequence so that the accent falls at the ends of the four-note groupings clarifies the patterned structure of *Bendrong*.

1 1131 1131 4543 2252 2252 2252 4543 1131
1131 1131 4543 2252 2252 2252 4543 1131 [. . .]

I should emphasize that making this adjustment was easier said than done. More experienced students tried to explain it to me, but intellectual understanding did not seem to help me play correctly. It wasn't until my "aha" moment – in which a new, hitherto inscrutable way of interpreting the sounds I heard quite suddenly seemed natural – that it clicked.

Once I made this adjustment, many other baffling details of the music made sense. Some gamelan instruments are meant to provide rhythmic markers; instruments such as *kenong* and *ketuk* (horizontal gong chimes) and *kempul* and *goong* (hanging gongs) articulate the metrical units outlined above. Each instrument's name evokes the distinctive timbre and pitch of its sound (*nong, tuk, pul,* and *goong* respectively). In most pieces, they provide parts that interlock, giving each rhythmic pulse a character all its own and articulating the larger rhythmic units (*tuk-pul-tuk-nong*). They began to serve as helpful aural signposts once I knew how to listen.

Learning to hear and feel gamelan music as end-weighted proved to be an invaluable experience while conducting fieldwork in West Java. It enabled me to engage with the music I studied with my Sundanese consultants in a manner more similar to their own, and facilitated my "musical discourse" with them. More importantly, it provided me with a real-world example of how a single musical surface might be interpreted in wildly different ways.

Gamelan ensembles in higher education

Almost from their beginnings, Hood-inspired university-based study groups transcended their original function as the musical equivalent of language labs. Hood did not envision these ensembles, peopled with novice musicians, mounting performances or becoming equivalent to school orchestras, bands, and chamber music ensembles. Ricardo Trimillos points out that examining "a forty-year trajectory" of world music ensembles "reveals development from a single academic rationale to multiple ones," ranging from an adjunct to scholarly inquiry, through a site for consolidating and expressing ethnic identities, to a purely aesthetic activity.[5]

University gamelan ensembles in particular have become quite visible on college campuses across the United States. Few would dispute that the instruments provide "glittering sonic and visual symbols of 'the Other,' "[6] Commentators cite a variety of qualities of gamelan music to explain its special appeal to Westerners. For one thing, given some relatively simple techniques, students are able to participate in full-fledged music making without rigorous training.[7] A general increased awareness of percussion instruments in the second half of the twentieth century was another factor.[8] This was only one aspect of gamelan music that happened to coincide with modern Western aesthetics. Cooper and Tenzer also note that, "[t]he rich textures and rhythmic complexity was what we loved about Balinese gamelan."[9] Others have characterized it as a "kind of phase music" *à la* Steve Reich.[10] Gamelan music also appeals to orientalist fantasies: emphasis on the natural qualities of the sounds – as evocative of "moonlight and flowing water"[11] – or on the imagined ancientness of the instruments.

Many of these reasons for gamelan's popularity illustrate my opening assertion about the non-universality of music. In Java and Bali, gamelan music is neither ancient and anonymous, nor static and unchanging – in its history or performance. Nor is gamelan music any easier to play than any other music. These notions are the result of misunderstanding and mistranslations – imagining familiar meanings where none exist.[12] Ironically, then, it would appear that university gamelan ensembles are in some ways diametrically opposed to my own ethnographic goals in that they reinforce, rather than challenge, oversimplified notions of music as a universal language.

Gamelan ensembles as a site for research

One way to disabuse students of universalizing assumptions is to try to instill in them something approaching an insider's engagement with the musical materials. I have experimented with different approaches for introducing unfamiliar concepts to the students – and my observations of how they come to grips with them intellectually and viscerally have honed my own ethnographic translations.

Let us return to the notion of end-weightedness. One approach that seems to click is to have students imitate the sounds of the marking instruments by singing *tuk-pul-tuk-nong* while they play their parts; this approach surreptitiously leads them to think of end-weighted patterns – it is hard to accent the insignificant *tuk* sound – without warning them that the rhythmic sensibility is different.

Another approach that helps to reorient students' rhythmic sensibilities is to teach them to sing a song first. I had always regarded Sundanese singing as difficult; not only is the language hard to pronounce, but understanding the relationship between the tunes and accompanying gamelan parts can be elusive at first. While gamelan music tends to be rhythmically regular, with a preponderance of simple duple meters and symmetrical phrase structures, songs often have more complex phrase structures and elastic rhythms that intertwine with gamelan accompaniment. At some point, however, it occurred to me that this fear of song was an artifact of my own mental blocks. The average Sundanese person – even one who is not schooled in gamelan performance – is quite capable of singing well-known songs and marking the moments when *goong* and *kenong* strokes coincide with the melody. The ends of the song phrases coincide with the ends of the gamelan phrases, but the beginnings do not.

It turns out that the average college student, too, can learn a catchy tune relatively easily, and quickly learn to mark the ends of the phrases with the appropriate instruments. Doing so builds an understanding of end-weightedness, because the notes fall at the ends of the sentences, quite literally like "punctuation," and predisposes them to reorient their musical thinking in that direction.

This experience also illustrates how I – a non-Sundanese individual who, despite many years of intensive engagement with Sundanese music, is most definitely an outsider – can gain new ethnographic insights by teaching neophytes who also are outsiders. This successful teaching technique exposed for me a bit of ethnographic wisdom: approaching Sundanese vocal music as end-weighted helps to make sense of its characteristic fluidity. This insight hitherto was opaque to me because of my own blinders, and was brought to my attention only because of the different experiences and competencies of the students.

Gamelan ensembles as research

University gamelan ensembles make a contribution to the training of researchers. My involvement with gamelan ensembles over the years has given me a measure of bimusicality that has enriched my ethnographic investigations. They also constitute a site for research; the gamelan performance classes I teach, in addition to providing a means to initiate a new generation of bimusical ethnographers, serve as a unique laboratory for my ongoing ethnographic investigation of gamelan music. Teaching gamelan performance generates important insights regarding insiders' and outsiders' conceptions and perceptions of the music's structures, and these insights have enriched my vocabulary for describing, analyzing, and translating these differences.

But the gamelan ensembles I teach are more than laboratories. In recent years, I have come to regard them as a kind of ethnography in and of themselves. If the goal of my ethnomusicological work is a translation of the meanings of gamelan music that are not immediately apparent to my fellow Westerners, what better way to do it than to actually empower them to make music – to learn to produce and consume those meanings for and by themselves? What better way to drive home the variability of human behaviors and engender a healthy skepticism of their own complacent reactions than to allow them to experience the results of a different system of behavior? Given the rarefied audience for scholarly writing, I would not be surprised to learn that more people have tasted the fruits of my ethnographic research through participation in my ensembles than through reading.

Hood's idea of developing "bimusicality" as a means toward true understanding of other musics is an admirable, but lofty, goal. Even truly bilingual individuals have accents that profoundly color their speech; it is not difficult to detect the American accent in even the best Western gamelan performances. Yet, less-than-perfect language skills are sufficient to enable discourse that might otherwise be impossible; my students are now capable of engaging in musical discourse with people from the other side of the world – what more can we ask from ethnographic research?

Notes

1. From "Ancient Spanish Ballads," Henry Wadsworth Longfellow, *The Prose Works of Henry Wadsworth Longfellow*, 2 vols (Boston, MA, and New York: Houghton, Mifflin, 1886), vol. 1.
2. Mantle Hood, "The Challenge of 'Bi-Musicality,'" *Ethnomusicology* 4, no. 2 (1960): 56.
3. Ibid., *The Ethnomusicologist* (New York: McGraw-Hill, 1971), 372.
4. Paul J. Revitt, "The Institute of Ethnomusicology at U.C.L.A.," *College Music Symposium* 2 (1962).
5. Ricardo Trimillos, "Subject, Object, and the Ethnomusicology Ensemble: The Ethnomusicological 'We' and 'Them,'" in *Performing Ethnomusicology: Teaching and Representation in World Music Ensembles*, ed. Ted Solís (Berkeley: University of California, 2004), 25.
6. Ted Solís, "Introduction: Teaching What Cannot Be Taught: An Optimistic Overview," in *Performing Ethnomusicology: Teaching and Representation in World Music Ensembles*, ed. Ted Solís (Berkeley: University of California, 2004), 7.
7. *See* Neil Sorrell, "Gamelan: Occident or Accident?," *The Musical Times* 133, no. 1788 (1992): 68; Hardja Susilo, "Sumbangan Para Sarjana Dalam Pengembangan Karawitan Di Amerika Serikat," in *Studi Gamelan Jawa Di Luar Negeri*, ed. Soedarsono, Retna Astuti, and I. W. Pantja Sunjata (Yogyakarta: Proyek

Penelitian dan Pengkajian Kebudayaan Nusantara (Javanologi), 1985), 30; R. Anderson Sutton, "Studi Kesenian Jawa Di Amerika Serikat," in *Studi Gamelan Jawa Di Luar Negeri*, ed. Soedarsono, Retna Stutui, and I. W. Pantja Sunjata (Yogyakarta: Proyek Penelitian dan Pengkajian Kebudayaan Nusantara (Javanologi), 1985), 6.

8. Sorrell, "Gamelan: Occident or Accident?," 67.
9. Rachel Cooper and Michael Tenzer, "Sekar Jaya: Learning from Traditional Bali," *Ear* 8, no. 4 (1983): 12.
10. Marc Perlman, "Some Reflections on New American Gamelan Music," *Ear* 8, no. 4 (1983): 4.
11. Mantle Hood and Hardja Susilo, *Music of the Venerable Dark Cloud: Introduction, Commentary, and Analyses*, ed. Mantle Hood and I. E. Records (Los Angeles: Institute of Ethnomusicology, UCLA, 1967), n.p.
12. One important aspect of gamelan music that does translate well into an American context: making music is an enjoyable social activity that emphasizes the group over the individual. *See* Sutton, "Studi Kesenian Jawa Di Amerika Serikat," 1–11; Perlman, "Some Reflections on New American Gamelan Music," 4–5.

27

Performative and Pedagogical Interventions: Embodying Whiteness as Cultural Critique

John T. Warren

When I was a doctoral student, a favorite professor of mine was famous for her first day creations – wonderful, evocative, embodied experiences that captured the course goals, the type of pedagogy we could expect, the very essence of what was to come. These varied from a tea party in feminist pedagogy to production worthy performances of Charles Dickens; each first day captured the imagination, left us awed at what had just happened and eager for the next class. Magic was spun in those classrooms. Since my own studenting days, I have longed to be able to do that kind of work – I want people to leave my classes talking of the power and wonder. If imitation is the greatest form of flattery, I have tried to create homages that she would be proud of. I think of such opportunities as true performative pedagogy – truly enfleshed moments where knowing, learning, and discovering are not just head games, but fully embodied and action oriented.

In this chapter, I describe a performative pedagogy that takes seriously the potential of bodies, of politics, of the radical potential created when lives meet in the context of performance – a kind of performance as research for the classroom. When working with students and teachers, I often talk about performative pedagogy as an intervention, a momentary and processual break in the expected, in the status quo, in the normalized glaze that covers and drenches inequalities and power with a sugar-sweet coating that denies the bitterness that underlies systemic oppression. These interventions are disruptive, are cracks in the allure, are leaks in the sedimented ways of seeing that, over time, can decay and fold into themselves. Performative pedagogy has the potential to expose power; without its protective invisibility, power can be seen as the artifice that it is. I teach culture and communication courses at a large midwestern university, often using literature in whiteness/cultural studies as a way to study and intervene in dominant ways of understanding the world. While the study and critique of racism has been the major site of my teaching, I believe that the power of performance, as I hope to show here, works across many content and disciplinary areas.

Performative pedagogy and whiteness have been linked since the inception of this new movement in "whiteness studies," a generic (but perhaps misleading and premature) term for the collective scholarship that began in the late 1980s and early 1990s which sought to deconstruct whiteness (by white people) as the center of power in (mostly) US culture. This link began with Peggy McIntosh's working essay on white privilege and her struggle with getting students to see oppression's concomitant effect: unearned advantage.[1] Asking students to create their own list of privileges, a common practice in teaching her essay, moves one toward a performative politics. Research on the pedagogy of the stage has also been offered, trying to foster hope in sketching and critiquing whiteness through theatre.[2] In each, the performance impresses upon the audience a critique of whiteness and a public pedagogy of sorts that seeks to replace normalized messages of dominance with images of hope and possibility. Recently, Leda Cooks and Jennifer Simpson's edited collection brings together a powerful variety of different pieces on classroom interventions in racist logics.[3] In this chapter, I seek to build from an essay I co-authored for that collection, which considers parody as a tactical response to the recurring presence of whiteness.[4] This vision of performative pedagogy is an enfleshed mode of inquiry – by engaging in creative parodies of whiteness, students and teachers come to see the possibilities of intervention made possible through embodied classroom engagements.

Parody and whiteness: performance as interventionist research

Judith Butler provides a groundbreaking treatise on rethinking those aspects of a cultural self that often go under-examined or altogether unquestioned, in which she makes the following summative statement: "My argument is that there need not be a 'doer behind the deed,' but that the 'doer' is variably constructed through the deed."[5] Butler reverses the normative ways of thinking about gender, undermining the idea of a stable gendered subject and crafting a vision of gender as a regulatory process that, by doing, creates that subject. In a sense, my male-ness is produced through my life experiences (from blue blankets in the hospital nursery to little league baseball games to performative expectations in high school locker rooms). These experiences are the conditions upon which my sense of self was generated (and, since such conditions are often so repeated as to appear normal, the creation of that self is obscured over time). What we see today as gender is, in actuality, the repeated, sedimented, and normalized façade of gender – each of us performing a self in service of the cultural expectations under which we all labor.

In our work, Amy Heuman and I craft a response to the entrenched nature of whiteness in our lives. We wanted to intervene, pedagogically, in a manner that met the complexity of the problem. To do this, we needed to examine the radical origins of whiteness and enter at the root of the problem; otherwise, we were in danger of leaving systems of power effectively in place. We turned to Butler's notion of parody as an intervention strategy, as a possible solution, embedded in research, that might deconstruct normalized constructions we were faced with. Butler has argued that "all signification takes place within the orbit of the compulsion to repeat; 'agency' then, is to be located within the possibility of a variation on that repetition. [. . .] it is only *within* the practices of repetitive signifying that a subversion of identity becomes possible" [emphasis in original].[6] A careful reading of Butler shows that in the very process of repetition one has the ability to act – it is in the process of remaking (again and again) of social systems that one might effectively intervene. For Butler, change lies in the "possibility of a variation on that repetition" – the thing done differently can alter the thing itself.

Amy and I discussed a number of parodic takes on whiteness that have been effective in our classrooms. One strategy is a first day activity I use with my students. I ask them (mostly advanced undergraduates or graduate students) to create group performances of whiteness based on their everyday experiences. I ask them to create images of whiteness with their bodies – what does it look like, sound like, feel like? As students build these collaborative performances, I ask them to exaggerate the details in performance. Whether they choose something benign like "whiteness is a lack of dance skills" or something like, "whiteness is terror," the outcome is useful and productive for the class.[7] The issues we need to talk about are immediately placed front and center; as the students craft such extreme performances they begin – in the relatively safe place of the classroom – to make fun of the elements of whiteness they are presenting. From overtly geeky representations of white folks' failed dancing to absurd examples of white surveillance, the performances allow for what Butler describes as "subversive laughter" or a laughter that "in the pastiche-effect of parodic practices in which the original, the authentic, and the real are themselves constituted as effects."[8] In the moment of laughter, we come to see whiteness as produced rather than natural. In that moment, whiteness is seen as façade.

What separates this way of engaging whiteness from others I have encountered is that the intervention is embedded not on the surface of whiteness production, but rather at its core. The power of whiteness comes from its ability to be unseen by those who wield it. Exposure is whiteness's greatest threat. Like the wizard in the classic film *Wizard of Oz*, the greatest danger the wizard faces is the pulling back of the curtain – such nakedness leaves him powerless. Whiteness, similarly, begins to wilt when seen. Such subversive performances break the spell, if for just a short time, and are powerful in their ability to interrupt the taken for granted, leaving all slightly changed as a result. In my classes, I use three parodic tactics that serve to challenge the normative role of whiteness in our culture:

First, I begin with interventionist performances that ask students to embody systems of power (much like the one above). Such efforts are powerful because they put abstracted ideas into the bodies of the students and constitute a kind of performed research. It is one thing to talk about whiteness as terror; it is quite another to embody it, to put such metaphors (and actual experiences) into the body. The enfleshing of ideologies, of systemic oppressions, and cultural norms – the taking up of these ideas in performance – leaves the traces of power in the tissue and consciousness of students.[9] Such experiences serve as a common integrating point of return; the performances are fodder for conversation throughout the semester and useful resources for our collective research.

Second, in every class, I use performativity as an analytical frame for understanding how identities come to be sedimented and normalized through daily, incessant acts. Certainly, the depth to which I go in each class changes depending on the audience and the course goals, but in some way we (eventually) come to performativity as an explanatory model for understanding how something that is created (like gender or race) comes to take on the status of the taken for granted (through repetition and normalization over time). For instance, in my sophomore level undergraduate course in communication theory, I try to explain performativity as the process of a snowball that, as it rolls down a hill, gathers momentum and mass. If you see the snowball only as a static object (as we tend to see something like gender or race), it appears to have substance; yet, seen in context, it only exists in as much as it continues to move down that hill (just as my whiteness is a product of multiple acts of reproduction governed by social norms and expectations). In other words, such snowballs are not accidents, they are products of time and individual acts of turning over itself along the track down the hill. Each time gender or race is affirmed and repeated, the snowball does another rotation. So gender, as an example, is repeated so often that its origins are obscured, hidden by the very process of its repetition. Since I use this way of thinking in my classes, I hope it continues to offer keys or ways of seeing identity that undermine its stability and allow for radically new ways of imagining for students.

Third, I often ask students to engage in a group performance project that allows them to work over time to create alliances with their classmates. I find this allows them to practice collaborative action, working with potentially different and diverse folks to create interventionist performances. As a rehearsal for change, these performances work to create the skill sets necessary for larger socially progressive networking. I also find these performances allow a kind of freedom to push the borders of their comfort zones in ways that individual projects do not. Since it is collaborative, responsibility is shared and allows for less personal and individual risk. In one particularly powerful course performance, I saw a diverse set of students, in the context of Toni Morrison's *The Bluest Eye*, create a collection of narratives that chart the Middle Passage, as African voices share what it might have been like to witness and experience the crossing of the ocean, the shift from subjects in the world to objects owned and sold by whites. The performance risked much, but provided a place for all to question how the present is performatively linked to the past (from the Middle Passage to slavery to Jim Crow).

These three tactics offer ways to enter, interrupt and subvert the normalcy of whiteness in the classroom through performance as research. A central commitment to this work is that interventionist pedagogies must come from a deep understanding of the issue, careful research into its history, and a dedication to possibilizing in ways that align with the angle of analysis.

Final thought

While my focus here has been primarily about how performed research should lead one's performative pedagogy, I end here by emphasizing how the performance work I do in the classroom builds and sustains my research. Performative pedagogy is not just a theory rich practice; it is a practice that builds theory. When my students engage whiteness, I find new and insightful ways of coming to know what whiteness is and what whiteness does. For example, in one performative exercise, students created a tug-of-war in which a white woman was being pulled between privilege and the desire for change. During one particular moment, the woman steps out and approaches the audience, asking each member of the class for help. "Will you help me? What should I do?" As audience members, we watched as the performer eventually ended back where she started, her arms outstretched as the double bind presumably continues. As the class began to unfold the complicated layers in the performance, we began to discuss the role of the audience and what it might mean that we each failed to offer assistance. As audience members, we knew our role: we watched and acted as if the struggle was not about us – it felt as if we would be ruining the performance if we had spoken up. Translating this performance to whiteness in the world, we began to see how whiteness is sustained through our inaction – one of the performances of whiteness in daily life is to witness without intervention. In this way, whiteness is affirmed by its continued unquestioned status.

As I close, I would briefly return to my former professor's opening day performances, those acts of intervention in the everyday, mundane happening of the classroom. Like the kinds of performative interruptions I have tried to sketch out here, those transformative moments alter how one sees the classroom. In those moments, we imagine our lives, our interactions, and our possibilities differently. I may not do opening days like that mentor of mine, but she gave me a desire to see what is and seek new ways of imagining what might be. This is the power of performative pedagogy as intervention.

Notes

1. Peggy McIntosh, "White Privilege and Male Privilege: A Personal Account of Coming to See Correspondences through Work in Women's Studies," (Working Paper No. 189, Wellesley College Center for Research on Women, 1988).

2. Shannon Jackson, "White Privilege and Pedagogy: Nadine Gordimer in Performance," *Theatre Topics* 7, no. 2 (1997); Anthony O'Brien, "Staging Whiteness: Beckett, Havel, Maponya," *Theatre Journal* 46, no. 1 (1994); John T. Warren and Amy K. Kilgard, "Staging *Stain Upon the Snow*: Performance as a Critical Enfleshment of Whiteness," *Text and Performance Quarterly* 21, no. 4 (2001).
3. Leda M. Cooks and Jennifer S. Simpson, eds, *Whiteness, Pedagogy, and Performance: Dis/Placing Race* (Lanham, MD: Lexington Books, 2007).
4. John T. Warren and Amy N. Heuman, "Performing Parody: Toward a Politics of Variation in Whiteness," in *Whiteness, Pedagogy, Performance: Dis/Placing Race*, ed. Leda M. Cooks and Jennifer Simpson (Lanham, MD: Lexington, 2007).
5. Judith Butler, *Gender Trouble: Feminism and the Subversion of Identity* (New York and London, Routledge, 1990), 142.
6. Ibid., 145.
7. hooks describes how whiteness came to represent those terrorizing faces that came into her childhood sanctuaries to brutalize and discipline Black people. bell hooks, "Representing Whiteness in the Black Imagination," in *Displacing Whiteness: Essays in Social and Cultural Criticism*, ed. Ruth Frankenberg (Durham, NC: Duke University Press, 1997), 170.
8. Butler, *Gender Trouble*, 146.
9. John T. Warren and Deanna L. Fassett, "Subverting Whiteness: Pedagogy at the Crossroads of Performance, Culture, and Politics," *Theatre Topics* 14, no. 2 (2004).

28
Open up the Box: Pedagogy, Action Research and Art

Kim Yasuda

In *Chaosmosis: An Ethicoaesthetic Paradigm*, Félix Guattari asks, "How do you render the school class as an artwork?"[1] With this, he challenges us to consider the teaching/learning environment and the work/activity that takes place there in order to realign pedagogy with practice and to question the way art is taught and generated within institutional frames.

I have taken up Guattari's question in a series of recent teaching experiments at the University of California, Santa Barbara through the implementation of action research (AR) assignments and courses in the arts.[2] My students and I have undertaken several multi-year projects: the design of public space for seasonal farm-worker housing in Oxnard; the retrofitting of a local bakery into a student-community gallery/café in downtown Isla Vista; and the renovation of used shipping containers as campus studio space and public sustainability labs. I will describe the latter in detail in this chapter as an example of opening up the box – literally opening up the shipping containers as well as the pedagogical and institutional frames of the art department and university.

These exploratory projects in class-as-artwork and exhibition-as-school constitute what I call "The Friday Academy" – a group of AR courses that function as temporary alternative teaching projects alongside and outside departmental curricula. With a "*public or perish*" mission "to remain within (the institution) without becoming (the institution)," the Friday Academy continually reinvents its pedagogy, promoting a sense of enterprise in the way art is taught and generated within the institutional frame, and values practice-based research. Straying from traditional studio arts training models, Friday Academy maintains an interdisciplinary assortment of academic refugees and community scholars who work in collaboration with one another as creative problem-solvers.

Using AR modalities, we address salient issues that emerge from a collective discourse rather than a disciplinary frame within the university classroom. We harness the resources and energy of a shared moment as well as its relevance and sense of urgency. Connecting our students and ourselves as teachers to the world at large, we become what Henry Giroux calls "border intellectuals, those who function in the space between high and popular culture; between the institution and the street; between the public and the private."[3] We negotiate and rearticulate the fluctuating role of a discipline such as art and its relevance within a broader cultural context. By opening up the teaching space for experiments in creative praxis, I have witnessed greater coherency between my sensibility and practice as an artist and my role as a professor – more than when I worked diligently to compartmentalize my scholarly profile into a separate set of obligations between teaching, research, and service. Likewise, these AR projects bring a level of coherence to the students that allows them to think and act beyond class-driven, instructor-led assignments toward sustained engagement in a process that challenges their initiative and fosters a sense of commitment to something that exceeds academic expectations.

Open Container: a case study

From its inception, the *Open Container* course had little semblance to the academic norm. It was described in the university catalogue as an intermediate 3-D studio art course to be team-taught by an interdisciplinary suite of colleagues, students, and community members.[4] Course content was improvised with(in) the community rather than developed in advance from a particular theoretical or aesthetic premise.

The idea for this course had been inspired, in part, by an exhibition in 2004 at the University Art Museum that featured the architectural installations of the New York based firm, LOT-ek. These included a full-scale, sleekly modified shipping container prototype that was called an MDU (modified dwelling unit), which was designed for a hypothetical lifestyle of compact mobility. The following year, we witnessed the devastation of hurricane Katrina and the inflated housing market in the Santa Barbara area. These circumstances further coincided with the loss of campus studio space for undergraduate art students. Working with shipping containers seemed to invite the possibility for testing both aesthetic and sustainable interventions in affordable housing as well as academic and community space. Within the fold of this research initiative, students became active participants in a public discourse by drawing upon their capacity as creative problem-solvers.

We initiated an undergraduate course that would use AR methods and the idea of a sculptural material response to the pressing need for the challenges in our community. A critical facet of AR practice is forging relationships and sustaining them in meaningful ways. Our early solicitations to local area businesses for donated containers resulted in an ongoing collaboration with business owner Jorgen Staal of J. Staal Storage Solutions. Staal offered his inventory with an enthusiastic challenge for students to figure out a renewed use for the containers. Staal remained an active participant throughout, attending critiques and offering his expertise and resources.

Shipping containers are an ample resource and constitute their own problem – boxes that "made the world smaller and the world economy bigger" in a mere 50 years are stacked up and idle at loading docks across the nation.[5] Cargo container walls have amassed to such heights that they impose premature sunsets on port-adjacent communities. Arriving in the United States full of outsourced product from China for our insatiable consumption, containers sit empty with nothing to fill them up; nothing makes their journey back across the ocean economically viable. The question looms – what can be done with these massive vessels, now worth less than the material it takes to make them?

In March 2006, J. Staal Storage Solutions delivered three 8′ × 8′ × 20′ shipping containers to the UCSB art department sculpture yard. Because of the project's scope, the course was extended beyond the ten-week cycle of the quarter system – from one term to the next over a period of two years. Each term of the project was numbered. Overall, more than 40 students participated along with contractors, roofers, vendors, and architects from the community who became involved along the way.

The first term, *Open Container I*, explored more "traditional" research into the container's 50-year history and its pervasive role in globalization, supply-chain economics, and distribution patterns of production and consumption. This also included research into new models for habitable space as well as presentations on architectural prefabrication. Paper and cardboard desktop exercises explored the formal aspects of "boxhood" and analyzed a multitude of hypothetical incarnations from the wildly imaginative (sensory deprivation tank plus bowling alley with exterior projection screen) to the sobering and practical (affordable housing, artist studio spaces, and panoptic prison complexes). The group prepared conceptual drawings and generated dialogue and critique toward the eventual formulation of a collective plan.

The next phase of research required that we begin to work directly with the boxes. Given the innovative research being done with container architecture by those trained in professional design fields, we continually asked ourselves, "What can artists bring to this discourse?" With the mix of creative ingenuity and hands on problem-solving capabilities, we accomplished something different than we might have, had we done it in the usual "college art classroom" way. In most cases, the students surpassed the teacher; they brought an unexpectedly unique set of skills to the table. With this collective enterprise, we navigated through the first year of *Open Container* to render a new working relationship with the community and an unusual work of art. The sheer physical presence of the research that had taken place surpassed anything resembling either a classroom assignment or more traditional student 3-D design work. Moreover, addressing the ubiquity of containers as material form and social phenomenon brought the art practice into an interdisciplinary realm. It was bigger than all of us.

Sustaining the life of a box

Numerous iterations of this initial project continue to circulate. *Open Container II* took place over the summer of 2006 when one container was moved from campus to the rear of the downtown Santa Barbara Museum of Art. Designed by graduate student Billy Hood as an expandable exhibition/drawing station with a six-seat, micro-cinema at its interior, the mobile gallery was placed adjacent to the museum space in conjunction with the exhibition, *Contemporary Chinese Photography*. Featuring works by our students as well as local and international artists, the UCSB satellite space functioned as a counter-dialogue to the museum exhibition, addressing relevant issues surrounding globalized labor and consumption. This mobile display unit now has a community life of its own, redesigned and deployed by a suite of new students and local artists, relocating itself behind high schools, libraries, and public spaces throughout Santa Barbara County. The parasitic nature of this exhibition space maintains its flexible institutional capacity to resonate with each site that it occupies.

Open Container III–IV (fall 2006/winter 2007) developed more slowly, as the project moved into painful lessons of practicality. To lend credence to the container's utilitarian application, we explored its capacity for habitation. We, as artists, tried to harness the imagination for our sculpture to function as a plausible dwelling unit as well as a strong artistic statement. This magnified expectations of the class to test both logistics and creative expression in the research outcome. The roofing and insulation phases took shape with sober pragmatism and less focus on the visual impact than artists are accustomed to making. Neither student nor teacher possessed the appropriate skill set and volunteer tradesmen were called in from as far as San Antonio, Texas. Such resource partnerships prove critical to AR formations. Community experts brought relevant bodies of knowledge and complementary forms of expertise to the university research pool, linking practical with theoretical problem solving. Similarly, the project facilitated cross-institutional resource exchange, as materials were provided by both the campus and community.

In its final version, *Open Container V* picked up pace in spring 2007 with a roster of new students as it moved toward becoming a fully accessorized art studio day space. Three years after the LOT-ek exhibition, the modified container came full circle, back to the sculpture yard, and was showcased at our annual undergraduate art exhibition in conjunction with the University Art Museum. Although the unit still lacks running water, it functions with up-to-code electricity and serves well as a day studio environment for an array of students who have occupied the two-container unit over the past year. When our undergraduate studios were condemned in 2004, my students and

I began to imagine how these cheap, mobile prototypes could be located on the same west campus site to serve as temporary replacements. If we were to entertain these studio-classrooms as readymade alternatives, we could provide some of our best students with immediate access to clean, well-lit spaces, preserving the productivity of one of our department's most successful undergraduate research programs. Pitched as a faculty-student research initiative, the "Box Studio Village" has received conceptual approval by the governing campus bodies and will be further developed over the next year with ongoing support from Jorgen Staal and the University of California Institute for Research in the Arts (UCIRA).[6] Our partnership has subsidized the next set of design innovations through a national competition for container re-use as well as a campus hosted, multidisciplinary conference on the container's 50-year history.[7] The selected design is currently in development and will be constructed by community and campus volunteers on university property in spring of 2009. Our original, art student-designed unit, alongside this professional version, will form the first phase of a public-accessible, demonstration center for research in sustainable re-use and affordable housing. As a satellite hub of the University, it will continue to link the research mission of the campus with the broader needs and interests of the community.

Postscript

Two years have passed since this project began. To this day, the box is in the department sculpture yard. Most of the students who dedicated a significant portion of their academic lives to it have graduated and left town, further opening the question of stewardship. How is a sense of ownership and long-term commitment instilled within the circumstances of a transient student population? Where does a transformation take place? How does the collaborative model of practice-based research foster a sense of community beyond the life of a class project?

The negotiated space of collaboration was central to the evolution of many of the student artists. This essential component to any AR project is not encouraged in the studio arts training model, which tends to privilege originality over shared authorship. Working in a group is not easy for the formative artist who has yet to establish their individual identity and voice. Amid over-scheduled academic lives prescribed to become degreed in four years, our young students will compete with the multitude of BFA- and MFA-trained artists emerging from university art programs across the nation and now, across the globe. Most of these students will join the workforce of postgraduate artists currently employed outside of the profession for which they have trained. Giving up exercises in personal vision for the collective consensus fervently engaged a few of the students, but also lost the interest of many. For those that chose to commit to the citizenship of a creative practice, a distinct breed of art student emerged who thought of themselves as part of an expanded community invested in an enterprise larger than the effort of one.[8]

The project provided an alternative way to teach and think about art, both its production and presentation. AR practice provides a place for the student, as well as the professor, to navigate his or her own role in the academic trajectory. Rather than perceiving of a practice as part of, or fitting into, pre-established institutions (museum, gallery, or university), we were able to envision ourselves as resource innovators of new venues in which research may be both produced and circulated. As in a work of art, the institutional formation itself may be conceived as a fluid project with malleable social and material relations.

When I started *Open Container*, I envisioned a project that would serve as a legacy of long-term investment within the academy and in the larger community. In retrospect, some of this was accomplished, while posing other challenges to the formula for engaged scholarship. For example, I have gone on to generate a number of other practice as research demonstrations with my students, setting up conditions that linger through and between the quarter system in the outside world, piggybacking onto existing structures on and off campus. We have undertaken renovations of local businesses and provided public art for seasonal farm-worker housing in collaboration with a non-profit housing developer. We have undertaken experiments in collaboration with other campus players including the library's *Campus Reads* program, the Student Health Office Drug and Alcohol Program, and the Associated Students Committee on a Safe Halloween. In every case, the transformative moment occurs most powerfully within the community my students and I serve, while leaving questions about how students are changed by the process and in what ways transformation occurs.

As a case study in form and function, and an ongoing example of AR in the arts, *Open Container* provides relevant data to inform future versions of teaching and learning outside the box. Recently, it served as the model project for the *2006 UCIRA Open Classroom Challenge Grants*, which were awarded to four UC campuses for innovative curricular research initiatives.

Through this work, we learn that the teacher–student relationship is not hierarchical – rather, it is a lateral and flexible dynamic of mutual engagement. AR work also demonstrates that the traditional research university is not necessarily positioned to facilitate the best and most relevant learning experience for our young students in today's world. As faculty, we must continually reinvent the system while remaining within it, figuring out ways to overcome its tendency to replicate itself and the kind of scholar it produces.[9]

Notes

1. Félix Guattari, *Chaosmosis: An Ethicoaesthetic Paradigm* (Bloomington: Indiana University Press, 1995), 134.
2. *See* Chapter 15 "Action Research" in the Cartographies section of this volume.

3. Henry A. Giroux, "Borderline Artists, Cultural Workers, and the Crisis of Democracy," in *The Artist in Society: Rights, Roles, and Responsibilities*, ed. Carol Becker and Ann Wiens (Chicago: New Art Examiner Press, 1995), 5.
4. I taught the project for two academic quarters with Dick Hebdige and Robert Wesheler. Student project leaders included Billy Hood, Koji Tanaka, Brooke Devenney, and Yak Yu.
5. For images of Open Container, go to www.arts.ucsb.edu/faculty/yasuda
6. The UCIRA is a multi-campus research unit promoting arts innovation across the ten UC campuses. Along with Dick Hebdige, I serve as its Co-Director.
7. The national competition was announced at the November 2007 UCSB conference, "The Traveling Box: Containers as Global Icons of Our Era." The conference brought together more than 25 interdisciplinary scholars to discuss the economic, social, and cultural impact of containerization. *See* www.ihc.ucsb.edu/containers/boxconference_schedule.html.
8. One student, Dominic Lang, went on to the graduate architecture program at Tulane University to work in New Orleans on emergency housing.
9. For images of *Open Container*, go to www.arts.ucsb.edu/faculty/yasuda.

29

Searching for Spalding Gray: PAR Pedagogy, an Undergraduate Ensemble, and "The Edinburgh Project"

Rosemary Malague

Searching for Spalding Gray is the title of an ensemble-devised piece created in Theatre Arts 275: The Edinburgh Project, an undergraduate course conducted at the University of Pennsylvania in 2004. The project was performed at the Edinburgh Festival Fringe later that year; its inspiration was the disappearance and death of the American performer, Spalding Gray. Gray once described the artistic process he used to create the autobiographical monologues that made him famous: "The first [stage] is trying to figure out what I'm thinking about and talking about by actually doing it."[1] Of the resulting performance pieces, Gray concluded: "I've come to know my life through the telling of it."[2] Gray intuitively and intellectually embraced the fundamental principle of performance as research: *knowing* is the result of *doing*. Director Tina Landau theorizes that kind of practice when she explicates the strategies of Anne Bogart's SITI Company: "*Composition* is a method for creating new work. It is an alternative method of writing. Rather than being alone in a room with a computer, Composition is writing with a group of people, on their feet."[3] In this chapter, I explore the notion of "performance as research" within a college curriculum, documenting the processes that led to *Searching for Spalding Gray*. What can be discovered through *doing, telling,* and "*writing on one's feet?*"

The course description for Theatre Arts 275 in the University of Pennsylvania's *Academic Bulletin and Course Register* (2004–6) reads:

> This course will combine an intensive practical and intellectual investigation of some area of the making of theatre: performance techniques, theatrical styles, a particular period of theatre history. One section of 275 in the spring will include six Theatre Arts students who will have been selected by the Theatre Arts faculty to participate in the annual "Edinburgh Project," and will form an ensemble to mount a production that will travel to the Edinburgh Festival Fringe in August.

The course topic varies depending on the instructor's areas of interest and expertise. Prior to *Searching for Spalding Gray*, I had taught the Edinburgh Project twice, once as "Twentieth-Century Acting Theory and Practice," facilitating an expressionist, cross-gendered production of Sophie Treadwell's *Machinal* (1997); and later as "Acting American Drama: Players and Playwrights," which produced a traditionally minimalist "Two by Thornton Wilder": *The Happy Journey to Trenton and Camden* and *Pullman Car Hiawatha* (2000).

In 2004, however, I began the course without a formal "topic." I was intrigued by the notion of venturing away from conventional scripts – the only kind of directing I'd done – and facilitating the creation of an original work. I wanted to stage a site-specific performance in an Edinburgh cemetery. Whether through folly or foresight, I was "haunted" by images of the city's burial grounds. I thought a cemetery would make an incredibly evocative theatrical site; an outdoor location might also draw passers-by and would be very much in the Fringe spirit. I had little experience in devising theatre so I began without really knowing what I was doing. To my surprise, I discovered that this approach provided my students – and me – with a rich opportunity to explore performance as research. The following is a narrative of the twists, turns, bumps, and detours we took together on the road to Edinburgh.

Searching...

With "cemetery" as my *idée fixe*, I announced my "plan" to the class and assigned Tina Landau's essay, "Source Work, the Viewpoints and Composition: What Are They?" (1995). Strictly speaking, these three rehearsal strategies are interconnected and used by practitioners with specialized training. The concepts of "Source-work" and "Composition," however, seemed accessible points-of-departure. Landau writes: "*Source-work* is a series of activities done at the beginning of the rehearsal process to get in touch – both intellectually and emotionally, both individually and collectively – with 'the source' from which you are working."[4] As example, she continues:

> Anne asked everyone to come in on the second day with a list or presentation which answered the question "What is German?" She was not interested in academic research, which would bring us to the material from our heads, but in subjective responses which would bring us to the material from our hearts and imaginations: Our preconceptions, our prejudices, our fantasies, our own memories and histories and culture.[5]

Following these fragments of Bogart's model, I asked students to respond to the idea "cemetery" or to the topic "life and death." I was interested, however, in having the ensemble respond not only subjectively but *also* with "academic" research, believing both would contribute to the creative process.

Working with these prompts, students brought in an array of sources: music, poems, images, and plays. In the beginning, these emerged mostly through free-association and prior knowledge; later, they were located through focused research. We listened to Nick Drake, the singer-songwriter who had committed suicide as a young man in 1974. Students identified "cemetery plays": the graveyard scene in *Hamlet*; Tony Kushner's *Six Playwrights Bury a Seventh*; Edgar Lee Masters's *Spoon River Anthology*; and the third act of Thornton Wilder's *Our Town*. We heard reports on death rituals from ancient times to the present, including Mexico's Day of the Dead; presentations on environmental and site-specific theatre; we saw photographs and learned the histories of the Edinburgh cemeteries we sought to use.

"Composition" was the means by which we began a performative exploration of the material that was generated through the Source-work. Landau writes:

> *Compositions* are assignments we give to the company to have them create short, specific theatre pieces addressing a particular aspect of the work [...] during the Source-work period of a rehearsal to engage the collaborators in the process of generating their own work around the source. The assignment will usually include an overall intention or structure as well as a substantial list of ingredients which must be included in the piece.[6]

In one assignment, derived directly from Landau's essay, I asked two pairs and one trio of students to create five-minute performances with these components: a beginning, middle, and end; at least one object; music; movement; and some element drawn from *Spoon River*. I suggested that they might incorporate sources they had already brought to class – music, poetry, and so on – or new inspirations. Their instruction was to feel free to do *anything*; the pieces could be literal or abstract. Throughout the start of the semester, the ensemble engaged in a mixture of research, conversation, presentation – and performance *as* research – by implementing our own interpretive versions of "Source-work" and "Composition."

Spalding...

Coinciding exactly with the start of the semester was the disappearance of Spalding Gray. On the very evening that class began, I first heard he was missing. Gray's disappearance became my preoccupation – one that ran parallel to my *occupation*, the running of the class – until the two inevitably intersected. I had been spellbound by the power of Gray's storytelling and had seen many of his performances: I was a fan. Like many of his fans, I became involved in the story of his life through his monologues. When he vanished, we were suspended mid-story.

Gray had long meditated on suicide in public and private. He had apparently made previous attempts, and evidence pointed to his having jumped off the Staten Island Ferry. For what seemed the longest time, there was no resolution. The semester continued. I had already planned for my class to view the 1988 Lincoln Center production of *Our Town*, featuring Gray in the role of the Stage Manager. Suddenly, connections were everywhere: *Our Town* led to *Monster in a Box*, with Gray's reminiscences of playing the Stage Manager; recollections of his mother's suicide; and reflections on death. Slowly but surely, Gray became part of our "Source-work." On 7 March, during spring break, Gray's body washed up along the East River.

By then, I had also made the unhappy discovery that, despite my dogged determination (culminating in a phone conversation with the director of cemeteries and crematoria for the City of Edinburgh), we were *not* going to receive permission to perform in a cemetery. I must credit my colleague, Cary Mazer, for his supportive suggestion. As I discussed *both* of my preoccupations with him – the Edinburgh Project and Spalding Gray – Cary pointed to what seems obvious in retrospect. "I think you *have* a performance piece," he said. "You should make a piece about Spalding Gray." Unmoored from my original idea (however vague), I felt I did not know what to do – until I realized that we had already begun the process. The class had read, reflected and, most importantly, experimented performatively on the themes of life, death, and places of "final rest." Our "Source-work" exercises, along with tragic happenstance, had prepared us for our project.

Once the class embraced the notion of shifting our focus to Spalding Gray, assignments included continuing to collect new source materials: articles on his disappearance and death; interviews with Gray, his family, and friends; reviews of his work; and an open invitation for student ensemble members to pursue whatever interested them. Some of the research took a conventional form, but always with the intent that the material would ultimately be embodied. Students examined the history of Gray's work with the Performance Group and his collaborations with Liz LeCompte and the Wooster Group; the models described by David Savran served as one inspiration.[7] "I think of myself as a collagist," Gray once said, considering his *oeuvre*.[8] Creating a "collage" also made sense to us, so we continued searching for its various pieces.

One key piece came from the decision that each student would create and perform a short autobiographical monologue in the spirit – if not the style – of Spalding Gray. These were not to be imitative pieces but variations on a form. I stressed that they need not be intensely personal or revealing – each actor should find his or her *own* way of telling a story from his or her own life. Gray's own methodology was to "write" by speaking. His monologues were recorded and transcribed – but not written by hand. Members of the ensemble were encouraged to work in this way, and the individuality of each of their stories and the variety among them was both delightful and moving.

I am certain that through the *performance* of these original pieces, students learned much more about the form than they would have through simply reading, writing – or watching.

Searching for Spalding Gray

We ultimately arrived at the title *Searching for Spalding Gray* because so much of the semester had been about *searching* – for a cemetery, for a project, for sources, and watching others, from afar, search for Spalding Gray. We had also been inspired by one of the first in-depth investigations of Gray's disappearance entitled "Searching for Spalding."[9] The following is an outline of the project's final form.

A small television sat on the downstage left corner of the stage for the duration of the piece, which began with two brief video clips from *Monster in a Box*, selected to introduce Gray to those who didn't know his work and to evoke memories for those who did. The clips were also chosen for their mixed tone: both humorous and dark, they included tragicomic elements characteristic of Gray's work; we attempted to adopt such a mixed tone within our homage. CNN's Anderson Cooper then interrupts, appearing on screen to announce Gray's disappearance.

Gray performed his monologues wearing a plaid shirt, seated behind a desk, with a notebook and glass of water as his only props. In our scenic elements, as with the monologue assignment, we endeavored to evoke but not imitate his style. The actors wore plaid shirts but each was unique. We used six "snack tables" as small desks, echoing Gray's own sets and affording us flexible staging. Seated at their "desks," the student ensemble enacted dialogue drawn from an online messageboard about Gray. Speaking as if in isolation, commiserating through cyberspace, actors gave voice to the writers: "I can't stop thinking about Spalding Gray," "Every day I check the paper for any word," "I pray this will turn into a new monologue," and so on. The CNN broadcast then resumed, briefly mentioning his 2001 car accident and the catastrophic injuries that led to his darkest depression and suicide.

The early "Source-work" and "Composition" exercises we had used in class led ensemble members to a real sense of freedom to create, as they responded emotionally, intuitively, and intellectually to our new "source": Spalding Gray. One student was drawn to a Bach violin concerto to which she choreographed what we called the "Spalding Dance." The entire ensemble performed abstracted versions of Gray's familiar gestures: the opening of the notebook; the sipping from the glass of water; the gripping of the desk; the intense lean forward; the serious expression; the bemused shrug. Through careful observation of Gray's body language and facial expressions, she had created a poignant evocation of Gray's essence and absence.

Another performer made a free-association, serio-comically connecting her own experience of sudden (but temporary) partial blindness to the eye-trouble related by Gray in his monologue, *Gray's Anatomy*.[10] She wrote the lyrics to a song entitled "Even Astronauts Get Depressed," and another performer wrote the music and choreographed the singing, dancing ensemble.

Other elements included a section we called "the Critics," in which students caricatured reviews of Gray's acting; explorations of a role Gray wanted to play, Konstantin in Chekhov's *The Seagull*; and a role he did play in the cemetery scene from *Our Town*. One student adapted and dramatized an essay by writer and filmmaker Hugo Perez, who was the last person to see Gray alive – on the Staten Island Ferry the night of his disappearance. Drawing on Gray's storytelling style, two performers told the tale in direct address to the audience.

The performance was "book-ended" by Anderson Cooper's return, announcing that Gray's body had been found. Then, to the eerie voice of Laurie Anderson performing "One White Whale," the cast began a somber funeral procession, each carrying a black umbrella, directly quoting the iconography of *Our Town* but underscored by postmodern music. I was drawn to Anderson because she had scored one of Gray's films and because the opening cry of her song sounded to me like keening and the lyrics struck me with a chill: "How to find you, maybe by your singing/ A Weird trail of notes in the water/ One white whale in all these oceans/ One white whale."[11]

The final words of *Searching for Spalding Gray* were taken from the online messageboard where we began; its bloggers had been writing for two months and now responded to Gray's death. These anonymous voices were once again embodied, expressing grief, anger, and loss. In closing, the cast performed a short reprise of the "Spalding Dance," ending with a lone woman, silently regarding the single empty chair and desk that remained on stage. She exits, and lights fade to black.

Performance as research and PAR pedagogy

I offer this chapter as one example of how a group of students (and a teacher), without experience in devising theatre, traveled unfamiliar terrain to reach a thrilling destination: a deeply fulfilling performance that was its own form of research. Though I had not begun by thinking of this project as *my* research, I have come to recognize what I learned from the experiment – an experiment that led me into new territory, where I led but also followed, where I entrusted (and I hope, empowered) student-performers to find their own way through the maze that is the creative process.[12]

Those of us who make theatre know the adage about reviews: if we believe the good ones, we must believe the bad ones, too. Our project had but one published notice, satisfying for what it recognized and understood. Sarah Jane Murray described our piece as "a sort of delayed public wake" for Gray;

she called the performances "brave, heartfelt and inspiring." She also wrote, I must note wryly here: "The piece takes on the feel of a lecture, with Gray as the regarded and spirited teacher."[13] Spalding Gray *was* our teacher; like him, we "figured out what [we were] thinking about and talking about by actually doing it." Our stumbling through the semester toward Scotland resulted in a performance process and product that truly constituted knowing through doing – and discovering through "searching."

Notes

1. Quoted in Richard Schechner, "My Art in Life: Interviewing Spalding Gray," *TDR: The Drama Review* 46, no. 4 (2002): 165.
2. Quoted in the 1 June 2006 performance of *Leftover Stories to Tell: A Tribute to Spalding Gray* at Performance Space 122 in New York. Devised and directed by Kathleen Russo and Lucy Sexton. Public performance at Performance Space 122, New York, 31 May – 4 June 2006. The piece was produced as *Spalding Gray: Stories Left to Tell* at the Minetta Lane Theatre in 2007. A published version is forthcoming from Dramatic Publishing.
3. Tina Landau, "Source-Work, the Viewpoints and Composition: What Are They?," in *Anne Bogart: Viewpoints*, ed. Michael Dixon and Joel Smith (Lyme, NH: Smith & Kraus, 1995), 27.
4. Ibid., 17.
5. Ibid., 18–19.
6. Ibid., 27.
7. David Savran, *Breaking the Rules: The Wooster Group* (New York: Theatre Communications Group, 1988).
8. Quoted in the 2006 performance at PS 122 in New York cited above.
9. Alex Williams, "Searching for Spalding," *New York Magazine*, 2 February 2004.
10. *Gray's Anatomy* is the title of Spalding Gray's performance that opened at Lincoln Center in 1993; published by Vintage Books in 1994.
11. Laurie Anderson, *One White Whale* (New York: Nonesuch Records, 2001), Sound Recording.
12. Thanks to the ensemble: Peter Bonilla, Maryanna Geller, Caroline Gordon, Sally Ollove, Billy Rosen, Lauren Sankovich, and Jenny Weiss.
13. Sarah Jane Murray, "Searching for Spalding Gray," *see* http://www.edinburghguide.com/festival/2004/fringe/review_theatre.php?page=s.

30
Valuing Performance/Practice as Academic Knowledge

Lynette Hunter

This discussion briefly outlines the practice as research developments in the PhD at the University of California at Davis (UC-Davis), and will reflect on the need for this area to grow in departments of theatre, dance, and performance studies more generally.

At UC-Davis we run a PhD with three streams:

- **Scholarly**: based on history, theory and criticism of Theatre, Dance, and Performance Studies
- **Practice as research**: integrating practice-based methodologies, laboratory exploration, and a range of traditional, tacit, and embodied knowledges, with history, theory, and criticism of Theatre, Dance, and Performance Studies
- **Professional**: for mature professionals with normally 30 years of experience who want to reflect on their experience and record the contributions that they have made to knowledge, or their creation of substantial new insights.

This chapter explores the basic thinking behind the second stream, practice as research (PaR), but first it should be said that this is but one in a number of approaches to the integration of practice and research. There are related areas called "research as practice" and "research through practice,"[1] to which I would add "practice through research," which this essay includes under the larger term PaR.

The PaR PhD is an exceptionally popular idea with students, but they are often surprised at how demanding it is. After all, the student has to work in both traditional scholarly written modes as well as produce new material in other media. However, the key aspect that distinguishes PaR is that is a practice-based element that needs to be accompanied by critical, historical, or theoretical reflections on, or analysis of, the practice itself.

Details of the PaR PhD at UC-Davis, Department of Theatre and Dance

The PaR PhD at UC-Davis requires two years of coursework, followed by roughly two years of dissertation development and production. PaR students take the same three core courses as the scholarly stream, and have to take at least one further full course in history, theory, and criticism. Their other courses may come from this field, or from the PhD and MFA courses available through a departmental liaison at UC-Davis called a Designated Emphasis (DE) in Studies in Performance and Practice. The DE allows students to work in a transdisciplinary way, which is important in performance studies and to a lesser extent in the studies of theatre and dance, because they are all relatively new disciplines. Performance studies also addresses knowledge shaped in new and continually changing ways by changes in technology and media, that makes its area of research necessarily transdisciplinary.

PaR students have one practice-based project in Year One, and at least one further in Year Two that usually forms the basis of an examination area in their Qualifying exams. They take these exams at the same time as other PhD students in the scholarly stream, in the first quarter of their third year, and usually develop one PaR area involving practice-based research through production, one PaR area exploring the historical, critical, or theoretical approaches to their work, and one scholarly area of study. They may prepare a further examination area if they wish. All PaR elements in the PhD have to be accompanied by a journal recording and exploring the process of production, a documentation of the performance of the production, and a written critique reflecting on or analyzing the production. The increasingly important area of documentation, theatography and archiving, is central to the activities of a PaR degree. This is partly because these students are constructing a discipline and laying down the basis for further research, and partly due to the ephemeral experience of much performance that is generating innovative and performative approaches to documentation.

During their final two years students develop a production in their chosen medium and write usually between 25,000 and 60,000 words of dissertation to support their project. The length is negotiable and dependent on their project. Typically PaR students and professors tend to overcompensate for the newness of the field by requiring and doing too much.[2] Close monitoring is needed to ensure parity with other students. The production element must be presented within the same quarter as their final oral presentation, which is based on the examiners experiencing their production first hand if possible, and if not, through a recording medium. All final productions have to submit documentation of performance, in line with all PaR elements: journal, record, and critique – in effect the final written dissertation represents the critique. The examiners also read and respond at the presentation to both the written and the practice-based forms of knowledge.

PaR projects at UC-Davis have included the writing and staging of an experimental play, video production, poetry reading performances, costume design archives, and visual arts installations, among others. But we are only in the early days of this program.

Why offer a Practice as Research PhD in Performance Studies (or indeed why take one)?

"Theory" is thought in process. Once it is regularized it becomes "rules." Practice as research, a version of "art as research," focuses on the process, on the production, on the edge of articulacy. As such it *is* theory. It becomes a pedagogical tool in itself as it defines an area which focuses on the self-conscious critique of thinking and knowing as logical, embodied, and, finally, rhetorical activities.

With the increasing focus on technologically mediated performance and the awareness that it changes very fast, people inside and outside the educational system have begun to understand that art is tied to modes of production that construct knowledge, create insight, and generate beliefs. It is the speed with which technology changes that has foregrounded the knowledge-making activity of performance arts in general. As such these arts fall clearly within the remit of research study and university-level education. In Western philosophy this has of course been understood for many years. Bergson and William James argued for the materiality of artistic production over a hundred years ago.[3] And many creative writers, for example Samuel Coleridge or John Keats from the early nineteenth century, worked against the idealist philosophies of their time to articulate the process of their poetics.[4]

In a university environment, especially in the humanities, artistic practice is often disparaged. In contrast, scientists usually "practice what they preach" – they work in laboratories trying to do the things that they eventually write about. They have legitimate and valued areas of research that are devoted to producing the machines that allow other scientists to carry out other experiments – one person's experiment may lead to a practical production, for example, a machine that allows one to study quarks, that becomes another's grounding context. Scientific knowledge grows in this way, and acknowledges different kinds of knowledge. The social sciences also do this to some extent, working for example with the effects on human beings of an implemented social policy. But the humanities, in which arts departments often reside, tend not to work with grounding contexts that are material and physically actual in the same way. They often work from documentation that has been taken as fixed and dependable – in the earlier part of the twentieth century they even used to go out of their way to discourage the understanding of process and how it could affect documentation, precisely because they were trying to emulate the apparent timelessness and dependability of "nature."[5]

At the same time, many of the paradigm shifts in the humanities have come from the foregrounding of particular methodologies and what they highlight in terms of process. For example, in literary studies, the New Critics (1950s) or the poststructuralists (1970s) each worked hard to articulate in detail the process by which texts came into being, even if many of their insights later became reduced to "rules" communicated by a recognized and ever-growing body of scholarship. The arts, on the other hand, have not usually articulated the processes by which artists work. The legacy of idealist philosophy has led in some cases to a naïve aestheticism based on transcendence and external power, often divine, as if the artist were especially inspired by something godly. However, as the arts have been incorporated into tertiary education, whose primary function is to critique (sometimes we need to remember that the word "academic" means "skeptic") the acquisition, form, application, and dissemination of knowledge, the arts have addressed this new challenge with varying degrees of responsiveness ranging from public critique sessions in the visual arts, to the development of new vocabularies for these new kinds of knowledge, and its own body of scholarship.

Choreography is an excellent example of a practice that has developed approaches to embodied knowledge that have been woven into scholarly work. The critiques of processes, from improvisation to the formal qualities of ballet, have begun to yield exceptionally sophisticated critical and theoretical thinking.[6] The writers working from choreography have taken their models from rhetoric, history, ethnography, phenomenology, gender studies, ethnicity studies, and many more. Because they are just on the edge of articulation of a previously unsaid process, these often hidebound discourses take on new life as they are tested in the field of the arts. Yet unless we make an effort, an effort that can be made through specifically foregrounding PaR, it will not be long until these new vocabularies for choreography cease to pay attention to their grounding assumptions.

Indeed the history of improvisation is interesting to follow at this juncture. Improv(ization) is often held, by the general public, to be a relatively undisciplined free-for-all, a method that encourages people to think that they are liberated when they are simply reproducing habitual actions. But analysis of improve demonstrates that artists cannot engage with it unless they are highly skilled, with exceptional disciplinary experience from which to draw.[7] Just as a writer cannot be expected to write well in any sustained way without years of training (we train people in reading and writing for at least ten years in most Western countries), neither can a performer expect to perform with the fluidity and confidence of a particular self (as opposed to the predictable quality of conventional characterization) without similar training and education. In parallel, it could be argued that if we had spent the last 300 years working on ways to talk about theatre, directing, designing, and acting rather than on writing, we would have just as sophisticated a critical vocabulary for the processes of performance as literary studies have for creative writing – and we might have just as reductive rules for criticism as well.

For the history of literary criticism also tells us something very important about the dangers of articulating process: the moment it is said, especially the moment it is said clearly, can often be the moment that it ceases to be a process. Part of the reason that some of the arts such as theatre have allowed themselves to sit on the edges of university education is that many practitioners, even though they may become educators, believe that writing about art destroys it. Yet there is nothing essentially reductive about writing, nor essentially transformative and engaging. Just so, there is nothing essentially transformative and engaging about art (and, of course, writing can be "art"), nor need it be reductive and habitual. As I have argued above, writing is simply where Western nations have put their critical energy for the last 300 years. But this is not to say that all artists should write about their work, only to suggest that writing about art is not necessarily going to undermine and destroy it.

One of the other reasons for the slow development of the serious study of theatre, acting, movement, and voice is the lack of appropriate experimental facilities. Partly as a result of a huge change in social policy toward funding for the arts, and a grudging acceptance of their importance, the arts were introduced into education through conservatory systems which trained for the profession and did not focus on research. When they were brought into university environments they were nearly always placed in the humanities, which have quite different requirements for scholarship, and one result was the loss of adequate resources. Hence educational work by the arts in most Western societies has simply not developed the kind of laboratory facilities that it needs to do serious research.

After several decades of existing in universities solely to provide more or less sophisticated cultural experiences for local students and faculty, it is beginning to be realized that the arts are no different to the sciences and humanities in their need for serious research to develop the leading edges of practice. Yet attempting to bring everyone back into a conservatory system is simply not appropriate for either the increasingly large number of people who would like to use the arts to open up their engagement with the world, or for the dedicated professional artist, who like a lawyer or doctor needs a research environment to develop new ideas. The commercial theatre, like the public hospital, is not often the best place to test out innovative and culturally challenging artistic forms.

A scientist usually works from a lab that is funded at least partly if not wholly from grants for which they, or they in a team, apply. The funds for such grants are difficult to get, but are plentiful compared to those in the arts, because there are often technological, medical, governmental applications that may result in commercial gain. In the arts it is notoriously difficult to assign value, let along evaluate the benefits of practice. However, history does teach us that the arts make a vital contribution to culture, society, and politics, and create benefits that reach far into education, health, and spiritual belief.

The sciences have gained recognition for their contributions to new knowledge over the past 150 years by being incorporated into the education system in order to carry out essential research. That research is by definition risky, and frequently does not achieve what it sets out to achieve – and while attempts are made to lessen risk, we all know that much can be learned from failed experiment. But the signal difference between the laboratory of the scientist and that of the artist, is that the former aims to construct a method by which anyone in the world can duplicate the experiment.[8] Whereas the potential instability and unrepeatability is something that scientists worry about, it is the lifeblood of arts practitioners. Innovative and experimental works in the arts do not usually make large sums of money – a fact attested to by the consistent need for arts funding over Western recorded history, whether it be a wealthy patron, a national policy or a rich foundation. Nevertheless, the arts as much as the sciences need space, time, and laboratory materials to do their work.

There is also a profoundly significant difference between the material of arts research and that of the sciences and humanities, in particular the material of theatre and dance. These two disciplinary areas cannot conduct arts research (as opposed to ethnography, technoculture, computing, history, textual/cultural studies, etc, *of* theatre and dance, and in contrast to Action Research) unless they have the opportunity to work with actual human bodies. Ideally, we need well-designed experimental spaces and the funds to bring in the people with whom we want collaboratively to experiment. Like the sciences, and unlike the humanities until very recently, the arts are strongly collaborative and this brings with it additional issues of space and personnel. Furthermore, we need the opportunity to develop this experimental work onto the public stage. If the scientist realizes new scientific knowledge by testing it within the natural world, the arts realize the new knowledge developed in and through human bodies by testing it in the public arena of other human beings.

What we are constrained to do for the moment is organize temporary laboratory conditions in which people with extensive expertise can meet, work, experiment, test, develop, and coalesce, further contributions to knowledge in the area. In the past, something like these temporary laboratories has existed in workshops in large sophisticated urban places, to which like-minded researchers have gone to develop practice that occasionally results in a production – in other words in a publicly disseminated form. What is beginning to happen is a tertiary education culture of research conferences that takes the workshop idea and exposes it to the rigors appropriate to university study, if possible within an environment of actual production, or public performance. These are the experimental laboratories of current arts practices. They are temporary only partly because of the ephemeral nature of performance, and mainly because there are no existing facilities to enable serious collaborative research.

Practice as research is a new discipline that is emerging from concerned research in the performing arts, in an attempt to articulate the value of theatre, dance, and performance in general. If articulation is not attempted, it is not only difficult to put one's finger on value, but the activities can also be more easily disparaged and ignored by social, cultural, and political forces. This is disastrous in a time when artistic sponsorship, which has been necessary throughout the history of theatre and dance and is vital for current performance art, is in a moment of ambivalence: should the state support artistic practice? should big business support it? should private individuals support it? should communities support it? PaR is at the same time articulating value and attempting to keep process to the fore, to ensure that practice informs analysis, that productions inform critique and vice versa.

At present we can only hope that our ability to value practice is not compromised by our institutionalization. At UC-Davis students are expected to understand this paradox, this internal contradiction, and continually to engage with it. Indeed, I would argue that it is necessary to achieve some institutionalization and a great deal more administrative recognition, to ensure that PaR develops in a way that we want it to within North America. In other words, I would argue, in a way that is not a rule-bound activity, but a theoretical process continually exploring and engaging.

Notes

1. Angela Piccini, "An Historiographic Perspective on Practice as Research," *Studies in Theatre and Performance* 23, no. 3 (2003).
2. Robin Nelson and Stuart Andrews, "Practise as Research: Regulations, Protocols and Guidelines: A Short Report ..." Palatine, http://www.palatine.ac.uk/files/903.pdf.
3. Patricia Rae, *The Practical Muse: A Pragmatist Poetics in Hulme, Pounde, and Stevens* (London: Associated University Presses, 1997).
4. George Whalley, *The Poetic Process* (London: Routledge & Kegan Paul, 1953).
5. For example, the early drive of bibliographical theory in the twentieth century was motivated by this scientistic claim.
6. *See* Susan Foster, *Reading Dancing* (Berkeley: University of California Press, 1986), and later works.
7. A. Cooper Albright and D. Gere, eds, *Taken by Surprise* (Middletown, CT: Wesleyan University Press, 2003).
8. As an erstwhile biochemist, I know that something as random as room temperature can completely corrupt an experiment.

31
Performed Research: Audience as Investigator

Marilyn Arsem

In this chapter, I will discuss two performances that I designed as research projects. One is a metaphysical inquiry and the other responds to a Cold War military site. Both serve as examples of engaging the audience in acts of performed historical research. They are not rehearsed performances, and the outcomes are not determined in advance. The works are primarily dialogic in nature and are designed so that the initial audiences engage directly in the investigation, sometimes with me and sometimes independently. Gathering audience responses afterwards is critical to understanding how the works operate and what was learned. It also becomes crucial to the ongoing life of each performance, as secondary audiences engage with documentations of these initial acts and conversations.[1]

Re-search

Once I choose a topic, I cast my research nets widely, gathering as much information as I can, following as many different directions as arise, without judging what is important or useful. I pay attention to where my curiosity leads me. I keep notes on images, impressions, half-formed thoughts and ideas, anything that rises to the surface from written sources, conversations, or site visits. As I begin to contemplate different aspects of the topic, I take up the questions in other performance works, as an embodied form of research. For example, the project *Writing Ada* (described below) involved not only considerable archival research, but also various kinds of embodied research. Particularly, the 2005 performance of *undertow* at the International Congress of Performance Art in Valparaiso, Chile was propelled in part by thinking of Ada Shepard's disappearance at sea. In *undertow*, I rolled through seaweed for many hours until it completely encased my body. Since Ada's body was never recovered one might assume that she dissolved on the bottom of the ocean trapped in seaweed. *undertow* became a way of thinking – with my body – about the experience of losing one's sense of self, drifting (see Figure 7).

Figure 7 *undertow*, a durational performance by Marilyn Arsem.[2] Photograph by Sofia De Grenade. Reproduced by permission of the photographer

Works like *undertow* are simultaneously performances in their own right and at least partly "research" for another work.

I also delay the design while I am engaged in the initial research. This gives me time to: (1) find a structure related to the content; (2) identify the issues that become most compelling to me; (3) consider how they might be examined in live performance; and (4) pay attention to my own process. I remind myself that I can create a process rather than a product; this allows me to step away from the position of an authority delivering knowledge and instead to take on the role of facilitator in a conversation between myself and others.

Analysis: *Writing Ada*

This project grew out of a series of performances that I had created in the late 1980s and early 1990s, in which I examined women's historical labor in relation to images of women in New England cultural history. My research included an exploration of earlier generations in my own family and revealed familial connections to the Spiritualist movement of the mid-nineteenth and early twentieth centuries. On one side, I had an ancestor (Ada Shepard) who had served as a medium for Nathaniel Hawthorne's family, and on the other, a relative had been a scientist working with a group testing the powers of a

Boston-based psychic medium. Both questioned the possibility of real contact with the spirit world.

Considering the impact of science on religion, I became curious about the ways people in Boston in the mid-nineteenth century negotiated a collision of belief systems, and how it influenced their understanding of death and the afterlife. To narrow the inquiry, I focused on Ada. Given the similarities between us – both raised as Unitarians, college-educated, involved in liberal causes, and living and working as teachers in Boston – I trusted that interesting connections would emerge.

In a few weeks of library research, I understood that nothing I might script could approach the experience of directly engaging with the artifacts of the era: books, newspapers, letters, and journals. My haptic experience in the archive generated another kind of knowledge; as a result, I began to craft a performance related to the act of historical research.

Almost immediately my notions of the performance shifted from one that was designed as a theatrical presentation in a public art space to an event that would take place in private homes. Rather than attempting to recreate a particular historic period, or even scripting a comparison between two eras, I became more interested in another type of discourse with participants. I began to understand the enquiry as a desire to engage with others in a discussion about our own beliefs concerning the afterlife. The story of Ada Shepard became the device to generate that discussion.

This kind of discussion is more effective when situated in the private spaces of a home, within the context of a particular social circle, and in a more relaxed setting where people sit with each other around a table. Placing it in the home shifts the piece from a discursive lecture/demonstration to a site of shared storytelling, where people read aloud to each other and conduct experiments. As such, it resembles the daily life of a nineteenth-century home where people gathered in the parlor after dinner for conversation and shared amusements, tea, and dessert.

In advance of the evening, each participant receives a packet of historical texts to read. Each packet has a different focus. All of the texts are first-person accounts, and all were written or published in Ada's lifetime. In some cases, the materials come from books that she owned. Sources also include unpublished letters and journals, newspaper articles, pamphlets, sermons, conference papers, novels, and scientific studies. Depending upon the makeup of the group, the packets might focus on religion, feminism, education, or the history of nineteenth-century Boston. Materials may include:

- multiple accounts of Ada's séances with the Hawthorne family, gleaned from published accounts as well as letters and journals;
- excerpts from Elizabeth Stuart Phelps's *The Gates Ajar* (1868) – an account of heaven and one of the best-selling books of the late nineteenth century, and from Ada's letters concerning her father's existence in heaven;

- excerpts from Julia Ward Howe's collection of essays on higher education for women, *Sex and Education* (1874), including a chapter written by Ada, as well as newspaper debates on the psychological risks of education for women;
- mid-nineteenth-century accounts of experiments testing the validity of Spiritualism, conducted by doctors, judges, and professors at Harvard University;
- sermons and an autobiographical novel written by Ada Shepard's husband Henry Clay Badger (*Bethlehem or Border Lands of Faith*, 1895) as well as sermons written by the minister of her youth, which provide accounts of Unitarian thought on the afterlife;
- newspaper accounts of the search for Ada after her disappearance at sea and private correspondence between family members in the following months; and
- instructions on how to form spirit circles at home, from early books on Spiritualism, such as Robert Hare's *Experimental Investigation of the Spirit Manifestations* (1858).

When participants arrive, old books and photographs cover the table. The evening begins with participants sitting around the table and reporting on what they have read in their photocopied packets. They quickly realize that their materials overlap, and in some cases report different versions of the same events. The discussion about these connections happens spontaneously, as we pass the old photos and books around the table.

After initial discussion of the historical texts, questions relating to individual audience member's beliefs further expand the conversation. I might ask, what were our images of heaven as children? What were we taught about the afterlife? What do we think will happen to us after we die? Has anyone ever felt any special connection with someone who has died? Did anyone ever have an uncanny experience that couldn't be explained? How do we reconcile those experiences with rational thought?

As our conversation unfolds, I offer as example my own eerie connections with Ada. At the same time, I transform my physical presence in front of them. Adding layer after layer – chemise, drawers, corset, corset cover, underpetticoat, bustle, over-petticoat, underskirt, overskirt and bodice, my body is altered to resemble that of a woman of 1874. I don't perform *as* Ada, nevertheless my behavior is altered as the clothing restricts my breathing, holds my back rigid, and limits my movement.

And so they begin to see me, experience me, in a different light. This is underscored by a literal change in lighting: I turn off the electric lights and place a gas lamp on the table. Vision is affected as the flickering flame responds to our movement, our breathing. The atmosphere in the room changes, as the lamp burns oxygen and fumes rise and circulate in the air. We are altered physiologically – the chemical balance in our bodies is shifted.

We huddle around the table to see each other in the lamplight, as I read aloud instructions from a nineteenth-century manual on how to be a medium.[3] The last activity of the evening is an experiment in automatic writing.

The conversation with each group moves in different directions, depending on the combination of participants and their own interests. It is also impacted by the degree of openness and trust that participants bring to the project and that develops during the performance. It is necessary for the group to agree that they will not repeat elsewhere what others reveal about themselves during the evening, as we share experiences, beliefs, and fears that are generally held very privately. I have done *Writing Ada* with seven different groups, (and a few more in some test versions). Since the event unfolds in conversation with each group of participants and is driven by their own concerns and questions, the direction and content of each performance is quite different. Each performance might more accurately be considered a continuation of the work rather than a repetition or restaging.

Analysis: *recent: remote*

recent: remote, commissioned in 1999 by Seattle's Center on Contemporary Art, is an event that was created on the site of a former Nike missile military base, now part of Cougar Mountain Regional Park in Issaquah, Washington.

I was immediately intrigued by the secluded, fence-enclosed site and curious about its history. One of the first things that I encountered was a small cement booth that looked like a guardhouse. The words "Nike Base History" were painted above the door. Looking through the open doorway I discovered the booth was completely empty. It was a sign to me of our recent amnesia (particularly in 1999) concerning the Cold War. Overgrown paved roads led to nowhere, buildings had crumbled, fragments of cement and metal lay scattered throughout the woods. Silence pervaded.

My research materials included government archives, city archives, and interviews with local residents. These revealed a complex history of the location and its many uses: first by Native Americans, then as a coal-mining site, a missile base, a munitions company testing ground, a public school, and now a public park. I immediately encountered, not surprisingly, discrepancies in the records, often in reference to the same events. I became most interested in the ways that stories were created out of fragments of information. No one knew the whole history of the site, and everyone surmised or invented aspects of the stories that they did know. Overall, there was more that could *not* be known about the location than could be known. Despite all the available information, it was difficult to generate a comprehensive picture.

I soon realized this was the most interesting feature of the site, and so I designed the event so that the audience's task was to construct a history of the site, an effort parallel to my own process. I embedded additional clues and materials in the landscape that focused on the different histories of the site.

I buried old photographs from the location, of its mining and military days, so that they were only viewable through periscope windows placed in the ground. Surveillance cameras and tiny monitors were mounted in trees. Voices from the taped interviews I had conducted emerged from the bushes, triggered by movement. Nearly invisible historic texts were suspended in trees and scattered in fragments in building foundations. Ticking was heard from underground. A message in Russian, which could only be read from the air, was mowed into the meadow. Deep in the woods, a file cabinet overflowed with all the printed materials that I had gathered in my research. These included maps of coal seams in the area, construction plans for the Nike base, American and Soviet Union military maps from the 1950s and 1960s, Native American land claim cases, studies on the impact of the local mining industry, and military manuals. It also included more than 100 books, pamphlets, magazines, and newspaper articles published in the 1950s and 1960s on the Cold War.

recent: remote was open for two days and audiences came at different times. On arrival, audience members signed authorization forms and received individual packets with instructions and information about the site. Some were given cameras, each had a different map of the area, and all had field excavation forms to fill out. Their role was a combination of archeologist and undercover agent. Their task was to gather information about the on-site artifacts and to reconstruct its history based on those clues. But each person had different information about the history of the site. Each had also received secret instructions pertaining to surveillance activities.

Ten performers participated in the event. Each chose an aspect of the history of the location, and engaged in a related activity. These included setting up archaeological excavations; collecting and identifying plant and tree specimens; marking and cataloguing human-made objects on the site; conducting a site survey for a new housing development; making a documentary about the park; and maintaining security. When the audience-participants encountered these performers, they received an explanation of the site and its artifacts through the lens of the particular history on which the performer had focused.

The event branched out in multiple directions immediately. There were many incidents that occurred between individuals throughout the site, and no one could possibly see everything or encounter everyone. It was never clear which people were performers and which were audience members, as everyone had similar roles and engaged in similar activities. Each person's experience of the event was only a fragment of the whole. If someone had information in his or her packet on the mining history of the site, then the performance was read differently than it would be by someone who had information about the Cold War or Native American histories of the site. Additionally, cryptic messages were delivered in private, and conversations by other people were overheard. It was impossible to gain a comprehensive understanding of the site.

During the following week, a debriefing took place in Seattle that allowed us to record how people interpreted the event. It was also an opportunity for both the audience and the performers to discover each other's secret agendas, since no one knew the nature of another's covert activities. There were real surprises – deep cover missions – of which even I was not aware. The participants also talked about how the experience triggered their memories of the Cold War, and a particular kind of paranoia associated with that time. The group was a rare combination of people young and old, conservative and liberal that resulted in a very lively, and at times heated discussion. Did we "win" the Cold War? Was our behavior justified? Is the war still going on? Finally, it was a conversation about how we continually revise and reconstruct history based on fragments of information and the biases of the present.

This kind of dialogic performance can provide a very rich experience for the participants, depending on their level of engagement; a very different kind of relationship unfolds between the artist and the "consumer" of the work. The participants make the artwork *with* the artist; and time, commitment, and effort are required from all parties.

Documentation, dissemination, and secondary audiences

Because I work as a performance artist within the visual arts community and as a professor of performance art in a school of visual arts, there is an expectation of a simple and ocular final product. In the marketplace particularly, funders and curators often challenge me to defend the value of this kind of work, in relation to the small number of participants, and the impossibility of generating a concise and easily packaged conclusion from the process (intangible research outcomes).

These challenges have made me pay close attention to the documentation and dissemination of the performance/research, in order to configure a version of the enquiry that will provide another kind of experience for a secondary audience. These documents have included interviews with the participants, their writings after an event, as well as written, audio, video, and photographic recordings from the early stages of initial research through to, and after, the event(s). For example, secondary audiences for *recent: remote* experience the work through my public talks, anecdotally from participants, from a video about the project, or from written material distributed or on websites. I originally thought that the documentation of *Writing Ada* would be in book form, but at this point I assume it will be a box of letters/texts/photos, which would include both my original research as well as the participants' writings. In that way, engagement with it would replicate the experience of the work, where both performer and audience assemble fragments that are discovered in no predetermined sequence. Anyone who views the documentation will be invited to add their own writing on the questions that they have encountered while reading the materials in the box. With those additions, the work and the research continue.

Notes

1. When I speak of secondary audiences, I refer to people who encounter reports of the live work. The report might be in the form of oral history, written text, video, photos, artifacts, or relics. In other words, the performance has another life in other media; it might be relayed person to person, or more remotely via websites, publications, or exhibitions.
2. Performed in Ex-Frigorífico Empresa Portuaria Valparaíso at POST, at the 7ma Muestra Internacióna1 de Arte de Performance, 19 November 2005, as part of the 1er Congreso Internacióna1 de Arte de Performance, Valparaíso, Chile, presented by PerfoPuerto.org.
3. Allan Kardec, *Experimental Spiritism: Book on Mediums; or, Guide for Mediums and Invocators*, 5th edn (Boston: Colby & Rich, 1874).

32

Miss Translation USA Goes to Cuba: Performance as Research toward a Performative Ethnography

Shannon Rose Riley

For the last few years, I have conducted research in Cuba on a range of performance forms. In order to develop an approach to doing "fieldwork," I have drawn equally from my practice as a visual and performance artist and my work as a scholar in performance studies and critical theory.[1] Particularly, I have engaged a persona from my performance repertoire – the aging beauty queen, Miss-Translation USA[2] – in order to develop a method and theory for conducting what I call performative ethnography. This chapter consists of notes and reflections on this work, offering an analysis of a performance as research collaboration with the Cuban performer, Roxy Rojo, at the *Festival Nacíonal de Pequeño Formato* in Sta. Clara Cuba (2006)[3] and the kind of ethnographic practice that I am developing from that work. I conclude by outlining some research outcomes and situating the project in terms of performance ethnography and ethnodrama.[4]

Field research as performance: developing theory through practice

As I prepared for my first trip to Cuba, I was asked how I would account for my position in the "field." I replied, somewhat improvisationally, that I would entrust a performance persona like Miss Translation before I would rely on notions of personal identity to account for the embedded power relations, projections, and misunderstandings that inhere in conducting "fieldwork" in any kind of "contact zone."[5]

My response pointed partly to my frustration with ethnographies that position the participant-observer through autobiographical narrative and partly to my inclination to understand things through performance. I do not think that autobiographical narrative or positioning oneself in terms of class, nationality, race, gender, or ethnicity are sufficient strategies to account for the power dynamics at play when a privileged researcher from a "first world" nation-state engages with/in different cultural contexts. Such strategies often, if unwittingly, reinscribe a notion of the participant-observer that remains strongly rooted in liberal humanism and often ignore the power dynamics involved in intercultural contact. Although autobiographical strategies aim to question the authority of the observer and foreground her presence as an outsider, they nonetheless put forth a theory of subjectivity that often runs against contemporary theories of performativity.

As a result of my slightly flippant response, I began to consider what it might really mean to go to Cuba "as" Miss Translation. I felt, at least, that conducting ethnography from the slippery position of a performance persona would be more in keeping with performative theories of subjectivity. Moreover, the homophonic relationship between Miss Translation and acts of mistranslation opened up a fruitful gap: I began to theorize that it might position me within the dynamics of exemplar and error – acknowledging the privilege of engaging in certain kinds of research while keeping a spotlight on the invasive potential of outsider status. It might allow a kind of "fieldwork" that could foreground that interactions between cultures are largely predicated on misfires, mistellings, misinformation, and misrepresentation and problematize the idea of having access to another culture. Miss Translation USA might allow me to frame my position in Cuba within the particularly complex and troubled history of international contact between the United States and Cuba that I have explored elsewhere.[6]

My initial idea was to wear Miss Translation's pageant sash on a daily basis for the ten weeks I would be in Cuba.[7] I decided against this approach for various reasons. A Cuban colleague, living in the United States, warned me that wearing a sash that reads "USA" might put me, or Cubans associated with me, in harm's way, albeit for different reasons. He was concerned I might become vulnerable by advertising my national affiliation in the context of the almost 50-year-long US embargo with its travel bans, suspicion, strategic misinformation, and economic restrictions. Cuban nationals who might associate with me could possibly expose themselves to sanctions by the Cuban state for associating with an *estadounidense*, particularly someone perceived to be traveling in an official capacity. A pageant sash endows the wearer with the official status of a cultural emissary, very much in the manner of Austin's performative. Although my performance as Miss Translation relies upon the strategic misfire of being an imposter beauty queen, by performing in public space without a conceptual or theatrical frame, the sash threatens to be taken up as the legitimate marker of a quasi-official and specifically nationalist status. I also tried to avoid turning Cuba into a staging ground for my own agenda, especially as my research on the history of Cuba–US relations reveals this oft-repeated scenario (think of Teddy Roosevelt and his Rough Riders).

Instead, I began to outline a series of actions and tasks that could be conducted in a performative manner. These included photographic documentation of "mis-takes" – a framing up of the researcher's misunderstandings. The goal was to use the camera to document instances that revealed my lack of understanding and the projection of an outsider imaginary, rather than to capture images of Cuban culture *per se*. Meanwhile, I was introduced to Roxy Rojo, a prominent drag performer in Cuba,[8] and he invited me to collaborate on a performance with him the following year in his revue show at El Mejunje during the performance festival in Sta. Clara.[9]

The performance at *El Mejunje*

After sporadic correspondence, I returned the following year with a partly scripted, partly improvisational performance for Roxy and Miss Translation. The conceptual frame was that Miss Translation was the only contestant in the *2006 Miss Representations of the USA Beauty Pageant*. Miss Translation was to participate in three categories: the swimsuit competition (her one-piece vintage leopard swimsuit recalling the extended US cocktail hour of the 1950s), the interview with the pageant judge (to be performed by Roxy), and the talent category, in which Miss Translation would read an original fairy tale titled, *Cenicienta y las Juntas de Trabajadores (Cinderella and the Workers Union)*.

The swimsuit competition consisted of Miss Translation dancing a Fosse-style routine to a minute or so of "Whatever Lola Wants." The lyrics, though in English, get at something about the US imperialist project in Cuba since 1898: "I always get what I aim for and your heart and soul is what I came for." This was followed by the interview. At one point Rojo asks Miss Translation how old she is. The latter replies that she was born the same year as the Revolution. Roxy responds "ah, but you look very good." The audience laughs – the scene connects an image of aging beauty with the revolution itself, and was particularly poignant in 2006 because of Castro's failing health. Roxy concludes the interview by inviting the contestant to demonstrate her talent.

Miss Translation sits on a chair in the middle of the stage and begins to read her version of Cenicienta (Cinderella) – the very stuff of bourgeois, heteronormative gender roles – mistranslating it into a seemingly utopic communist fairy tale. In this version, Cenicienta becomes so efficient with her labor that she organizes the other workers in her neighborhood. They meet regularly in a local barn and soon decide that Cenicienta should attend one of the Prince's balls to convince him to implement land and labor reforms. One night, the fairy godfather of the revolution, the Future Ghost of Che, appears to the laborers.

At this moment in the story, the lights go out on stage. Miss Translation stands up on her chair, dons a military beret with a rhinestone star, is wrapped in a green toga in the manner of the Statue of Liberty and is handed a lit cigar, which she puffs in the darkness. When the lights go up Miss Translation has become the fairy godfather and the audience roars, some laughing rather nervously. Miss Translation/Che waves the cigar, and in good Marxist fashion, the materials appear for the workers to make Cenicienta's gown and slippers. S/he waves the cigar again and tells Cenicienta that a pumpkin-turned-1957 Chevrolet will pick her up the night of the party, but warns her to return before the stroke of midnight. Miss Translation removes the beret, hands off the cigar and green fabric, and sits back down.

The story continues: At the ball, Cenicienta is prevented from speaking to the Prince. At the stroke of midnight, she runs toward the Chevrolet, but drops the reform notes. Prince Charming finds them and begins his search for Cenicienta, but in order to look for her, he disguises himself as a revolutionary by dressing in green and growing a beard. He attends labor meetings in order to look for her – and his ideals begin to change. When he finds her, he reveals his disguise and the two are married. The prince redistributes the land in the kingdom, renounces his title, and they spend their time working to improve the community. The tale concludes: "Of course, in this utopia, everyone lives happily ever after. Or no? Perhaps it's only a fairy tale."

The piece was well received by the large audience. It was ambivalent enough that those who support Castro's government read it as a strange, but essentially optimistic communist narrative; more critical audience members thought it parodied the idea of communist utopia. I was humbled by the privilege of being able to perform in Cuba, knowing full well I would not be able to "do Che in drag" if I were a Cuban national. At the very least, the performance strategically and ironically foregrounds the violence of political collision between the United States and Cuba, rather than erasing it. It also mediates my position as a white female scholar/artist from the United States conducting research in Cuba. It foregrounds me in my own self-parody, self-critique as an Anglo-centered persona.

The day of the performance, Pedro and I rehearsed at El Menjunje. We decided that Roxy should mime the part of Cenicienta and another actor improvise as Prince Charming. This was not only a formal choice to create action on stage and supplement my static position as reader, but offered other possibilities. Roxy-as-Cenicienta brought the humor of her drag revue to the performance, helping to couch our critique sufficiently for the Cuban political climate, but at the same time making the story all the more radical. For, if Cenicienta represents Cuba and the possibility of Cuba's revolution, as many audience members assumed, then Rojo's mimicry questions what constitutes a national hero. Being gay was prosecutable under Cuban law until the early 1990s, and revolutionary masculinity, as exemplified in the homophobic figures of Che and Castro, has been framed as the normative model. In this context, Roxy-as-Cenicienta and Miss Translation-as-Che, dressed in the manner of the Statue of Liberty, become even more complicated.

At almost the last minute, we both agreed that whenever Miss Translation struggled with reading or mispronounced a word in Spanish,

Roxy-as-Cenicienta would correct her. These corrections escalated during the piece, from gentle smacks on the back to rather aggressive kicking. The audience loved the moments when Roxy, a kind of cult figure, displayed power by punishing and correcting the US beauty queen. In this sense, although Miss Translation has the voice (reads the tale), Roxy-as-Cenicienta has the power to punish Miss Translation's mispronunciations and mis-tellings. And although Miss Translation remains seated most of the time and does not "act" (except briefly as Che), Cenicienta and the Prince act and have the potential to disrupt the performance. At times, they had the audience in such fits of laughter it was difficult to hear the story above the noise.

PAR toward a performative ethnography

Anthropology and ethnography are often concerned with performance, from rituals and folklore to cultural performances. Moreover, there has been a relationship between performance and anthropology/ethnography since the work of Richard Schechner and Victor Turner. In *The Anthropology of Performance* (1986), Turner calls for a focus on the function of performance in research.[10] The same year, James Clifford, George Marcus, and others further shifted the position of the ethnographer from authoritarian observer to participant-observer. The participant-observer conducts research activities with an acknowledgement of the partiality of her knowledge, an awareness of the performative qualities of expressive speech and gesture, and an attention to the process of ethnography as an instance of "writing culture" rather than describing or discovering it.[11] Jim Mienczakowsi describes the "overall and rapidly expanding move towards ethnographic performance as a logical turn for a number of human disciplines in which *culture* is increasingly seen as performance and *performance texts* as being able to concretize experience."[12] Norman Denzin and others attend to the performance of the participant-observer in the research context, theorizing that the practice of ethnography needs to focus on what Joost Van Loon summarizes as "the performative aspects of sense-making" and the "auto/biographical aspects of our own involvement as 'ethnographers.'"[13] Van Loon claims that the emphasis on the performative in ethnography is drawn in part from Judith Butler's theory of performativity.[14]

Nonetheless, the analysis of performance and the performative in ethnography continues to focus on issues of writing, voice, and utterance. This is more in keeping with the linguistic legacy of the performative as developed by Austin than it is with Butler's notion of performativity, or the "quality of being created and sustained through repeated performance."[15] The focus in contemporary ethnography has been on: (1) the investigation of performance forms; (2) the performance of writing; (3) the use of autobiography as a way to acknowledge the researcher's own position in the work, and (4) the performative mediation of culture and meaning through language.

I argue that despite interests in performance and the performative nature of language, ethnography has not sufficiently explored the performativity of the researching subject in a context of radical difference. Where ethnography has turned to the production of performance, it has tended to replicate a focus on narrative and to maintain formal principles. This is the case with the approach known as ethnodrama, which is described as "performed ethnographic narrative," and aims to adhere "to the principles of a formal and recognizable ethnographic research methodology."[16] Moreover, ethnodrama has been developed primarily in mental health contexts and is similar to psychodrama.

I suggest that a performance as research approach to ethnography would draw more substantially from Butler's theory of performativity and alienate the role of the researcher in a Brechtian sense. This alienation would not be aimed toward others, but toward the researcher herself, highlighting her role as outsider-performer. Above all, she must never rely uncritically upon a naturalized (personal, authentic, autobiographical, affective, emotive) position in the research situation.

The goal is not to focus on a "self" that goes to Cuba and has experiences that are written into a research narrative or performed in a final staging as in the case of ethnodrama, but to develop a researcher-as-persona that emerges in the fieldwork context by means of a largely out-of-context performativity. If identity emerges through repetitive citational performance based on tropes that are culturally specific, what happens in the context of another culture? How are we "read" in that context and what are our citational strategies? How can we possibly account for the various uptakes, and so on? How can we account for our position in the work without reinscribing the problematic of liberal humanism? In this sense, Miss Translation USA does not "go to" Cuba; Miss Translation USA cannot be said to exist before or outside of Cuba. Whether I wear the pageant sash in public or not, I am still a radical outsider in Cuba and cannot pretend to be anything else; including autobiographical narrative in my "writing" hardly begins to unpack the various meanings and uptakes at play in the Cuba/US fieldwork scenario.

Performative ethnography aims to use performance, such as the collaboration with Roxy Rojo, as both a component of its research and an example of its "writing." The performance at El Mejunje is an embodied writing of the intercultural dynamic between Pedro and me, between Roxy and Miss Translation, between Cuba and the United States. It is a component of the research itself.

Performative ethnography is thus based on a theory and ethics of intercultural mistranslation, including the negative dialectic between an epitome (for example, a national beauty queen) and something amiss (mistranslated). It problematizes the idea of having access to another culture. In order to approach another culture, one needs to begin by unpacking the specific materialist history of representation and interaction between the researcher's primary culture and the site of her study. Only within the context of this embedded history can fieldwork be performed. In this sense, my performance persona, Miss Translation USA, an aging white American beauty queen, is a "model" ethnographer who highlights the history, violence, and desire of intercultural contact between the United States and Cuba. The performative ethnography itself aims to perform, rather than erase, these particularly heated relations.

Outcomes

Several research outcomes emerge through the work. First, is an awareness of the importance of using improvisation in theorizing and conducting performance as research. Second, there is the generation of data through two primary sources: through the performance itself and through the videotaped performance document. For example, because we continued to use improvisation within the context of the beauty pageant structure many actions in the live performance revealed information regarding the dynamics involved in Cuba–US relations. This was immediately clear in the escalation of the "corrections" described above. Although Pedro and I had agreed that he would correct me when I made errors in Spanish, we did not determine how he would correct me. That those corrections increased in violence throughout the piece recalls the embedded hostilities between the United States and Cuba. That escalation was immediately perceivable by both the performers and the audience.

Other discoveries were made only after repeated viewings of the video. For example, when the Future Ghost of Che tells Cenicienta that she must return to the car on time, Roxy-as-Cenicienta makes a gesture as if she is wiping her hands of the idea. The gesture is so dismissive that I could not understand it at first. However, recently Pedro and I spoke about how tired he is of the figure of Che as a revolutionary hero, given the man's position on homosexuality in Cuba. With this comment, I began to understand his dismissal of Che's commands during the performance. The video shows that this gesture is picked up by the audience – they begin to laugh immediately when he wipes his hands and shrugs his shoulders. It is a deeply coded gesture, as he is also dismissing the US woman who performs as Che.

Finally, there is the outcome of building community across the divide of embargo – I travel to the festival each year and continue to develop performance work with Pedro and others. I also serve as an unofficial archivist for the festival, documenting many of the performances and distributing digital images to the participants. By performing in Cuba I make my "self" vulnerable to Cuban reception and available for future projects. In this sense, I gain access as a kind of outsider/insider in the Cuban theatre community. Instead of documenting Cuban performance in my initial phase of fieldwork, I performed for a Cuban audience in a way that has changed my relationship in the field/to the field – many people involved in the Sta. Clara performance community (the initial site of my research) still refer to me as Miss Translation.

Notes

1. Portions of this chapter were presented in 2005 and 2006 for the UC Graduate Group in Practice as Research; in the Seminar for Performance as Research at the 2006 American Society of Theatre Research (ASTR) Conference, Chicago; and at the 2007 Latin American Studies Association (LASA) Biennial International Congress, Montreal, Canada.
2. Miss Translation was developed in 2001 as part of a conceptual project that explored the homophonic relationship between negative states or actions (miscommunication, misrepresentation, and so on) and the names of fictitious beauty queens (Miss Communication, Miss Representation, etc.).
3. Roxy Rojo is the performance persona of Cuban artist Pedro Manuel González Reinoso.
4. Some questions are outside the scope of this short chapter, but I would like to at least raise them here: why is it appropriate, even desirable, to use theory to initiate, supplement, or authorize creative work today even as it remains largely taboo to turn the tables and engage creative practice in the development of "academic" work? The complex circular relation between "theory" and "practice" involved in this project questions the rather simple notion that creative and academic work are entirely separate fields or functions. As others in this volume suggest, I find that the performance work often articulates something before I am able to put it into writing (*see also* Hunter, Chapter 34 in this volume). And if the processes of research and practice are in a more circular, mediated, and close-knit relation, then what are the implications for the ways we think about knowledge and disciplinary divisions?
5. This is Mary Louise Pratt's useful term. Mary Louise Pratt, *Imperial Eyes: Travel Writing and Transculturation* (London and New York: Routledge, 1992).
6. Shannon Rose Riley, *Imagi-Nations in Black and White: Cuba, Haiti, and the Performance of Difference in US National Projects, 1898–1940* (Ann Arbor: UMI/Proquest, 2006); ibid., "Racing the Nation: Cuba and Haiti in US Performance, 1898–1940," presented at the American Society of Theatre Research Conference (Phoenix, 2007).
7. The sash reads, "Miss Translation USA 2006" on the front and "mistranslation" on the back – in English.
8. Special thanks to Marc E. Blanchard for making the introduction.
9. The Cuban state presents small-format theatre festivals on an annual basis in both Havana and Sta. Clara. These are nation-wide events in which short works or works with only a few actors compete for prizes in various categories. Theatre groups come from all over the island to participate. The one in Sta. Clara is housed at El Mejunje, an open-air theatre space and community arts center known as a center for gay culture in Cuba.
10. Victor Turner, *The Anthropology of Performance* (New York: Performing Arts Journal Publications, 1986).
11. James Clifford, "Introduction: Partial Truths," in *Writing Culture: The Poetics and Politics of Ethnography*, ed. James Clifford and George E. Marcus (Berkeley and Los Angeles: University of California Press, 1986).

12. Jim Mienczakowski, "Ethnodrama: Performed Research – Limitations and Potential," in *Handbook of Ethnography*, ed. Paul Atkinson et al. (London, Thousand Oaks, CA, and New Delhi: Sage Publications, 2001), 469. See also Norman K. Denzin, *Performance Ethnography : Critical Pedagogy and the Politics of Culture* (Thousand Oaks, CA: Sage, 2003), 82–3.
13. Joost Van Loon, "Ethnography: A Critical Turn in Cultural Studies," in *Handbook of Ethnography*, ed. Paul Atkinson, et al. (London, Thousand Oaks, CA, and New Delhi: Sage Publications, 2001), 281–2.
14. Ibid., 282.
15. This tidy summary is from Marvin Carlson, *Performance: A Critical Introduction*, 2nd edn (New York: Routledge, 2004), 222.
16. Mienczakowski, "Ethnodrama: Performed Research," 470.

33
Acting (on) Our Own Discomforts: BLW's [Media] Performance as Research

Julie Wyman

BLW (**B**orcila **L**ewison **W**yman, "Be Like Water," or "BeLoW") is a process-oriented artist-activist collective, comprised of Rozalinda Borcila, Sarah Lewison, and myself. Since 2004, we have worked to develop an embodied mode of inquiry that uses performance to investigate political agency at a moment when oral competence and physical presence are being displaced by media forms. Our interactive performances take the form of re-speaking workshops and public meetings. The workshops invite audiences to perform, along with us, the memorization and public recitation of significant recordings in the history of radical media; public meetings convene audience and interested passers-by in urban locations that blend public and private space.

The performances survey existing power structures and our places within them. We use physical presence and the act of speaking about research media, culture, and the problem of the vanishing body, voice, and political agency in this age of corporatization and digitization – and then hope to intervene in this problematic by developing new capacities for dissent.

Our aesthetic is anti-spectacular rather than theatrical. Our pieces intentionally appear messy and "unfinished." They represent the effort and sometimes the failure of both performers and audience, as we grapple with the act of re-speaking militant words and confronting the invisible parameters of social control in public locations. Both performers and audience learn by intimately encountering the unfamiliar. We inhabit physical locations that are new and alien to us. When we re-speak, we learn and express words and sentiments that feel foreign and impossible.

We attempt to "cross-over" – to understand across the distance created by history, race, class, and social positioning – by occupying actual space or speaking others' words. Crossing-over does not produce a seamless, spectacular, or even convincing, performance. Instead, we reveal an awkwardness or discomfort with/in our subject matter. Our performances stand as an unlikely form of participant-observation; the successes and failures speak to issues of performativity, cross-identification, and media spectatorship.

This chapter reflects on the stakes of BLW's performance work, focusing on our practice of re-speaking and our public meeting process as forms of performance as research (PAR). I will describe some of the work and discuss some of the critical questions our practice raises. In one instance, the audience's refusal to re-speak a text sheds light on BLW's contribution to discourses about appropriation, identification, and media spectacle. In another instance, our ritualistic, repetitive routine of engaging with our site constitutes a performed research that materializes more than any amount of talking. For BLW, the performance event exists as a laboratory designed for the dynamic development of ideas, rather than the presentation of complete, polished material.

Pilot TV – Chicago, October 2004

In 2004, the Chicago-based conference, Pilot Television: Experimental Media for Feminist Trespass, assembled artists, performers, media-makers, and academics at a conference to address the question, "what would activist television look like today?" Responses ranged from a talk show for the politically depressed to a re-enactment of Eisenstein's *Battleship Potemkin*, performed by female-to-male transsexuals.

The conference opened with the screening of several archival recordings made by radical video collectives in the late 1960s and early 1970s. Recordings of militant speeches, protests, and man-on-the-street interviews bore witness to the sense of determined, radical leftist organization of that time. But the act of looking back from the political moment of 2004 was complicated. We felt inspired, but also daunted by the sense of agency and optimism that we saw in the speaking subjects and the video collectives' recording acts itself. From the standpoint of post-9/11 and the dissent-deterring Patriot Act, these recordings seem to fail to keep that which they preserve alive. Instead, they seem to evidence their moments' death – making radical speech seem further out of reach than ever. In a sense, the tapes revealed to us the failure of the medium, or at least the failure of passive, video-viewing, consumption of the past.

Ultimately, the conference – especially those opening videos – sparked the formation of BLW and our re-speaking practice. We wanted to use the complicated viewing experience we had had while watching the archival footage to address Pilot TV's central question, and decided to respond to the recordings by producing "live" versions. We aimed to intervene in the video medium by using our own bodies and voices to "play-back" the recorded speeches and we staged an unrehearsed performance that asked, "Why does it feel impossible to speak radically now? What happens to the speeches – what happens to us – when we re-speak their exact words here and now? Is there a way to close the gap between viewer and viewed – to mess up the tidiness of the usual video-viewing experience? What can the act of public presentation – subjecting material to the performer–audience interface – reveal about that material?"

In the performance three days later, we re-spoke what we considered the two most radical recordings of the collection: Queen Mother Moore's 1973 address to the male inmates at Greenhaven Prison, recorded by People's Communication Network; and Fred Hampton's 1968 interview, recorded by Videofreex. Moore's speech critiques the US justice system and advocates bearing arms against the police. For example, she relates how Marcus Garvey's followers used weapons to keep police away so Garvey could speak in public.[1] Fred Hampton's interview, recorded when he was under surveillance by ColntelPro prior to his murder, reports on the Black Panther Party's successes. He details social services such as the Breakfast for Children program and expresses, passionately, that the Panthers will never be defeated.

We delivered the speeches before an audience of mainly young white artists. Each of the three of us recited Moore's text and then we offered the audience an opportunity to read the words. Our method of re-speaking was intentionally un-theatrical. We did not "act" as Moore or Hampton, nor did we wear costumes. Our goal was not to perform "convincingly," but rather to use the means of performance as an inquiry into the meaning of these texts and the circumstances that allow their delivery.

As a form of performance as research, we found that re-speaking leads to various discoveries about radical speech, the transmission of experience through speech, and our historical moment's relationship to earlier time periods. The re-speaker is able to learn viscerally from the cadence, repetition, and vocabulary of the text as well as from any gaps or discomforts opened up by speaking the words aloud. For example, speaking Queen Mother Moore's words felt risky and complicated. We felt uncomfortable re-speaking her call to arms and affirming this advocacy of violence. As a group of three white women, we were also mindful of the potentially appropriative danger of speaking an African-American woman's words. We were aware of the shift in context, as the speech was originally intended for a primarily African-American audience. Our stance against representative theatricality, and our desire to speak these words without trained affectation, was, in some part, an attempt to acknowledge both the impossibility and the undesirability of taking on Moore's racial identity. Re-speaking suggests the possibility of connection and knowledge through difference rather than similarity. The experience of re-speaking is precisely *not* about identification, but about measuring the distance between us/her, then/now, and so on. Through re-speaking, we open up gaps between us, as performers, and the archived speaker(s).

For the audience too, performance is crucial to the learning or research process. In other words, both performing and witnessing an act of re-speaking produces new knowledge in a way that watching or listening to the recorded sources alone cannot. The audience not only hears the speech, but is also presented with a person laboring with remembering and speaking the text. The words and the performance layer on top of each other – as do the identities of the archived speaker and the re-speaker – creating a sort of palimpsest of information. The result is a sort of super- or meta-mediation of the text, its context, and our own political moment.

Another time and race: lessons in crossing over

Re-speaking is a fruitful PAR method because it brings a surprising array of knowledges and responses into discourse. For example, a group of participants at the 2007 UC Digital Arts Research Network conference expressed serious concerns about re-speaking Moore's words. On the strict grounds that only African-Americans should speak them – and to do otherwise would be an act of appropriation or worse, erasure – a group of artist-academics refused to participate in the workshop. "Just show us the video," they said. "Let Queen Mother Moore speak for herself." Another participant disagreed, noting that playing the video was no less problematic than re-speaking Moore's words. The video medium creates a safe, distanced position for its spectator and a clean, hierarchical division between the consuming viewer and the consumed object of view. Further, the video has already been framed, edited, and so on – usually by someone other than the speaker – further challenging simplistic claims to the agency of the archived speaking subject.

The idea that Moore's words about collective loss and disenfranchisement can only be claimed or understood by those in the same racial category seems preposterous and even pernicious in both its essentialism and exclusivity. If access to speak about, object to, or try to comprehend an experience of injustice is the private property of the victims, how can we hope for communication, comprehension, and identification across identity categories? Certainly, such essentialist notions of identity run counter to more performative theories – and re-speaking tends to expose performative aspects of speech and the speaking subject.

What is important here is that the gap opened up by the speaking bodies and the racial identity of the archived speaker reveals new knowledge. We became aware that cleanly separating our re-speaking practice from a legacy of white appropriation of black performance was impossible. Our attempt in this workshop, as we articulated at the event's end, was to stand against that legacy and explore our dis-identification or un-crossable distance from Moore's speech – both in the case of our racial identities and the speech's determined militancy. The performance and the conversation about identity, performance, race, and the video image that resulted allowed us to reconsider our re-speaking practice's contribution to a discussion about identity politics and the performance of identity. Within the context of PAR, the work provides a laboratory for exploring current concerns about power, identity politics, and the inhibitions and sense of resignation that such concerns sometimes spur.

BLW continues to acknowledge and discuss the potential dangers involved in re-speaking someone else's words. At the same time, we acknowledge that speeches like Moore's do speak across lines of race and time. Just as prominent as the racial difference between us and Moore, was the distance between historical moments that our performance negotiated. Our attraction to Moore's speech was both a connection to her words as well as a desire to better understand the gap between our political moment and hers. Moore's speech compelled us because of the way that it does speak effectively, across time and the bounds of identity politics-based categories, to our sense of urgency to learn to speak, organize, and resist in an era when free speech is threatened.

Shake it off: BLW's public meetings in Millennium Park

Chicago's Millennium Park, which opened in 2004, is meant to provide an oasis of comfort and refreshment at the heart of the city's downtown. Soothing, with its gently sloped paths through the botanical garden and lawns, and stimulating, with its interactive sculptures and Frank Gehry-designed curving reflective surfaces, the park introduces a twenty-first-century "human-scale" space into the center of Chicago's twentieth-century modern skyscrapers. The park is meant, above all, to be a place of pleasure and enjoyment. So why did we feel so uncomfortable there?

In July 2007, BLW announced a piece entitled "A Meeting is a Question Between." The work consisted of five consecutive days of public meetings in the Park and was part of an exhibition organized through the University of Illinois, Chicago by the collective, Feel Tank Chicago. The park intrigued us with its contradictory status as a site situated on land deeded, in perpetuity, to the people of Chicago, but occupied exclusively with corporate sponsorships. The Boeing Pavilion, the Chase Promenade, the AT&T Plaza, and the BP Bridge: nearly every concrete block and manicured green square proclaims its corporate sponsor with prominent signage. Millennium Park thus seemed a fitting site in which to continue our research on political agency and dissent under capitalism.

We wondered about our relationship, as park visitors, to a larger political and economic system in which the spectacle both overshadows and undermines our capacities to resist or dissent. What definition of citizen does this site produce? What is the larger function of the pleasure produced by this location? And what sort of connection between recreation/relaxation and capitalism does the site evidence? Is it possible to question, oppose, or alter the workings of this space from within the space itself?

For five days at lunchtime, BLW invited visitors to discuss the park. Sitting, standing, and sometimes moving around an easel-type meeting pad, we made presentations in which we offered information, raised questions, and reflected with our audience on the Park and its performativity.

The meetings began with a historical consideration of Millennium Park's land and the practice of privatization such as enclosure acts in England during the seventeenth to nineteenth centuries, which limited the commons and often allowed landowners to appropriate public lands. We then connected this history to the Park by showing the audience an 1836 map of Chicago, which designates this section of land as a common, to be held by the State. The map reads, "Public Ground. A common, to remain forever open, clear and free of any building or obstruction whatever."[2] Today, the land is nevertheless occupied by a privately owned railway station and underground parking garage. Only recently was the land above the garage rooftop donated back to the public, as Millennium Park.

The next phase of our meeting looked at how advertising by corporate sponsors in the Park masks various ways in which these corporations work actively, and in some cases violently, against the people. We began with the words "Public Park," written on the easel pad and soon changed it to read "Publicity Park." We asked our audience, "What, besides publicity, does corporate sponsorship of a promenade, plaza, or gallery in the park provide?" We shared research we had conducted prior to the meeting with our audience on, for example, Boeing's production of military technology that is geared toward killing more people for less money. We were joined at this point by a concerned citizen who told us to just have fun; the park is about having fun. When we asked how to do that he replied, "just shake it off." We tried to shake, as he demonstrated, but ended up dizzy and not feeling any better.

I made a presentation about the corpus of the corporation on the AT&T Plaza in front of Anish Kapoor's sculpture, *Cloud Gate*. The sculpture is a gigantic, bean-shaped fun-house mirror that allows viewers to see themselves in all sorts of shapes and configurations. I begin by noting that the corporation is a bodiless person, but if it did have a body, perhaps it would look like Kapoor's sculpture. Not only does it seem to be a body engaged in an act of navel gazing, it is also a giant mirror that allows its viewers to gaze at themselves. A self-ish sculpture. We began to use performance as a way to explore these ideas. A few of us got into the shape of the bean, trying to copy its capacity to mirror and multiply. Later that day, BLW invited more than 40 participants from the 4th July Parade of the Politically Depressed (an event from Feel Tank's Pathogeographies exhibition) – all dressed in pajamas with messed-up hair, streaked mascara, and signs saying things like "Depressed? It might be political" – to the Kapoor site for a performance interaction. Once there, we held up our signs and looked at ourselves. Rather than individual bodies, a group body was mirrored back. With our pajamas, signs, and streaked faces mirrored in the funhouse surface, we became in a sense the corporate corpus – a group of politically depressed consumers reflected on the funhouse surface of public land. Ultimately, we were told to stand back from the sculpture because we were getting in the way of people having fun [...].

Conclusion

BLW uses performance to research political agency in the new millennium. Our work is also activist in nature; our intention is not only to map out a lack of political engagement, but also to devise modes of resistance. The re-speaking workshops open up a laboratory for looking at issues of race, gender, nationalism, and also provide audiences with access to recordings of radical history. Our workshops acknowledge the inevitable separations created by history and lines of racial, gendered, and socio-economic identity, but also ask audiences to forgo the position of safe consumption that these separations allow. It is in this "crossing over" that we discover more about the text and its relevance today. Our public meeting practice used the act of "performing" to share with park-goers some of the psycho/pathogeography of Millennium Park. The format of the meeting – designed to present, discuss, and evolve raw ideas, utilized the vulnerable state of performing to survey the functionality of pleasure and resistance in the park. In both of these modes, the act of performing allows us first, to research actively, and second, to bridge the gap between scholarship and activism – moving beyond the analytic to the synthetic and experiential. The performance-audience dynamic and the vulnerable act of self-presentation provide a model, perhaps a microcosm, of broader cultural processes of political action, activity, and passivity. Our modes of performance, therefore, allow us to understand and intervene further, in social performativity.

Notes

1. Garvey founded the United Negro Improvement Association in 1914.
2. *Canal Commissioner Map,* (1836). Available from http://www.encyclopedia.chicagohistory.org/pages/300062.html (accessed 5 September 2008).

34
Theory/Practice as Research: Explorations, Questions and Suggestions

Lynette Hunter

What is Research?

- Searching for things that happen in the search, making presence like skilled improvisation.
- Searching more, finding copiousness, context, or the installation.
- Searching again, with the repetition necessary for commodity markets, seeking representation and desire.

These are three quite different searches, all of which can be made by research.

What is Art?
Short for "articulating"?
Or what happens before articulation, the attempts at saying what hasn't been "said" before.

Do they fit together? It depends on whether you think the "things" are there before you find them, *or* are constituted by the process of the search.

I argue: Art that searches and constitutes things through the process of articulating is research. This, by the way, is also a good definition for science.

As a practitioner I have developed performance art into a medium for exploring theory. Over the last 15 years there have been eight productions that I have toured to different places, that use either one-person scripts or performance installation scores to explore theoretical discourse.[1] The discursive fields of critical, cultural, and performance theory are usually mediated through conference papers or publication, in other words nearly always through the written word of formal argumentative structure. But by definition any *theory* is always on the edge of articulation because it's trying to talk about things that haven't been said before. This makes it difficult to communicate theory within the conventional discourse of conference and publication, and we have probably all had the experience of drifting off at precisely the point where we know that the speaker or writer is trying to say something interesting and it's coming out obscure.

The performance art that I have developed simply extends the logic of this situation by foregrounding the difficult to articulate theoretical engagements and materializing them in an embodied way through performance or installation. Responses to these performances lead me to deduce that people at the very least have more fun with theory this way, possibly because they are suddenly given license to give themselves up to the energy of the moment rather than pin themselves into a pen of decontextualized logic, what Lorraine Code calls "*s* knows that *p*."[2]

But the "not-said" is more complicated than this. One of the central concerns of current performance theory is that powerful political structures train us to listen, speak, see, and feel in specific ways, ways that often exclude the possibility of recognizing people and actions, animals and environments, that do not fit the parameters of those structures. Much political theory is concerned with how to encourage the voicing of those unrecognized, and hence usually unvalued lives and landscapes.[3] However, while sometimes the not-said is the result of this kind of obscuring and erasing discourse so that it is at times a not-heard, at others the not-said is the result of something that has not yet come into being, and at still other times the not-said is in effect the not-yet-said, simply existence for which we have not yet found a vocabulary – the whole field of tacit knowledge in philosophy is concerned with this last kind.

My performance work is concerned with all three: it usually employs analogical not-yet-said material from the realms of tacit knowledge such as food or make-up or sound (all of which are currently acquiring vocabularies, which have not yet been formalized). It is, as I note above, concerned with the precipitous edge of not-said concepts and feelings that theory attempts to bring into voice and being through the process of its performance. And it is fundamentally concerned with the rhetorics available to those not-heard by socio-political power that attempt to silence, exclude, erase, and obscure – willingly or by default.

The performances started by extending theoretical concepts, such as "Is the concept of the tragic subject appropriate for First Nations (Native American) culture?" (1995),[4] or "Can a Man be a Woman?" (1994),[5] or "Does theoretical discourse erase the socio-political immediacies of women's daily lives?" (1996).[6] More recently the performances coalesce out of various installation energies and give rise to theoretical speculation. I did not anticipate this change from performance used to elaborate research to performance as research. It has just happened. The work now is more improvisatory in the sense of Simone Forti.[7] It depends on past experience being re-membered through the body placed in particular socio-geographical locations. For example, *Bodies in Trouble* (1997) generated a series of theoretical discussions among other people that had nothing to do with me.[8] It was a nine-hour site-particular piece on women's labor, the textured and subtle modes of communication made by simple daily tasks, and the corollary enclosure in political structures. The piece transformed my practice away from demonstrative embodiment to embodiment that happens in the moment.

The move began to shift the authority for significance from me as "critic" to my body as medium, and this also had implications for how the work was evaluated because the scholar-performer was no longer in charge of the meaning. I was no longer interpreting work and "making theoretical statements," but providing the ground for others to think theoretically. Yes, I constructed the parameters for this, but the thinking that went into it was not generically communicated. "Bodies in Trouble" was a score for movement rather than a script with words, although for those with keen eyes words did emerge from the tapestry that was built on the hanging steel rods that were part of the installation.

The academy values words above all else. Even scientists need to put their experiments and interventions into nature into verbal form before they "count" as research. And despite the 18-month gestation of *Bodies in Trouble*, the exceptional amount of research and planning that went into it, and the exacting performative elements, even my own attempt with the help of a colleague to verbalize some of the implications after the fact fell far short of a script.[9] At the same time the piece did generate critical and theoretical response in others and therefore some peer-review that could be recognized by the university system. But the experience underlined the fact that the academy can only value certain kinds of words, words that are certain, or have a set of *a priori* assumptions.

Yet words are rhetorically just as capable of embodiment and process as bodies. Scores and scripts are equally as capable of being generic as they are of being in process. For reasons dictated by the medium, I returned to a combination of words and body in *Coming to the End of the Line* (1999), but found the work prompted into poetics when it was transformed into a written version.[10] The process marked a significant turn toward poetics in my critical writing as a whole, and to experimentation with words, voice, and text in performance. *Domestic Tragedy in the Household* (2003) is a media collage of webcam, present voice, recorded voice, present performance, and video performance that I am currently putting onto the page with a combination of Photoshop and text. And *De-Scripting Performance* (2007), which began as a poetic lecture (1995), is a rewritten version of a long poem by bp Nichol Photoshopped into the watermark of the paper which carries my critical interchange with Nichol's poem itself.[11] Significantly, the experimental poetics of these two pieces leaves them "unreadable" in academic terms – in the latter case literally so with the publisher having received "returned" copies on the basis of "misprinting."

The most recent work I've pursued focuses on collaboration, displacing the critical and authorial voice even further from the scholar. Despite the extensive work over the last decade or more on collaboration as an important rhetorical strategy for artistic, social, political, and cultural work, it is not a recognized form at university level. In science, groups of people work together in what has been defined as collective research, and famously the individuals are all named, often in order of proportion of contribution, at the head of each paper. Not only has this kind of research proved difficult to justify in the arts and humanities, but genuine collaboration in which there is no predictable outcome is virtually unheard of anywhere in the institution.

The piece *How Analytical Thought Stops You Thinking*, first performed as an unconventional keynote at the University of Birmingham conference, Beyond the Book (2007), brought together analytical bibliography with book construction, art-book concepts, and poetry. The hour-long scored, but unscripted piece explores the interface between hypertextual performance devices and hot-metal press printing and binding, and involves complex hand-press printing, and then the interactive (that is, involving all people present) staging of the folding, gathering, stabbing, and binding activities of traditional book production, and a radical intervention into cutting that impacts on the way the text can be read. Participants are invited to read from the final cut-books, and the piece becomes increasingly vocal. Its collaborative rhetoric made it quite unpredictable, and it remains to be seen whether the university will accept it as research.

It also remains to be seen whether or not I write it into words, and what kind of words those might be. The impetus was a long poem I wrote in 1994, which may constitute any evident need for a script. Yet it is fascinating how easily I get caught back into attempting to "write" the performed. Perhaps the crisis this generates has led to more performative writing. It is curious how many theoreticians have turned to the performance of the written, as if dwelling in the process of the words materializes their engagement with the world, brings the not-heard, the not-said, and the not-yet-said into being.

Connections

Connected to these developments toward collaboration and the dynamic between the verbal and movement in my own theory/performance work, is a deep-running interest in situated textuality and the philosophical thought around situated knowledge and tacit knowledge. This epistemological research focuses on collective and collaborative modes of knowing, and my own development of the field has been the study of how these modes of knowing engage in the process of articulating and form communities around both similarity and difference.

Much of my research has been concerned with how writing and reading and the verbal arts in general promote these different kinds of textuality, especially in poetics and the rhetoric that bridges the poetics into daily life. This concern developed in a particular direction: from 1994 I began to study the effects of dynamic movement in the Daoist tradition, studying a classical Chinese system brought to Europe from northeastern China in the 1920s. What fascinated me about the material was that the physical vocabulary was taught tacitly, and that the knowledge required collaborative strategies to articulate. In fact, the knowledge did not become knowledge until it was articulated and that articulation was often haptic, somatic, or kinetic. Yet collective articulation, an ability to copy the movements, even embody them, without attention to their situated qualities, fell short of Daoist knowledge, specifically of any subtle understanding of the paradoxes. Collaborative engagement that was attentive to responding to the situated was much more likely to generate a textuality that was involved in articulating those ideas of complementarity, negotiation, reciprocity, and friendship that also emerge consistently in the philosophy of situated knowledge.

After five or six years' training in this Daoist system and learning its vocabulary it is expected that students go out and teach others, at the same time as maintaining their own study. Applications abound in the educational field, in the areas of mental and physical health, in sport, community centers, and leisure centers, as well as in business and industry. My own work in the UK was primarily with women's refugee groups from Cyprus and from Somalia, and in the United States I have worked with an alternative high school, in what a number of people would probably call Action Research. Over a series of projects, the work has concentrated on using the skills as a basis for communication between myself and the groups, between the group and the local society, among members of the group, and enabled Access accreditation[12] for further education. These projects generated a few scholarly publications, but the focus for me was on practicing my research into the system in different situations and exploring the kinds of textuality that were constructed.

Movement is just as likely as any other medium to use strategies that construct both positive and negative rhetorics. For example, a physical image can be liberatory, exploratory, evasive, and so on. People working with the theories of Augusto Boal can sometimes forget that the strategies he tried to foster among people unrepresented, or rarely/poorly represented, in mainstream power could also be used among the powerful to bolster their position. In other words, there is an ethics in tacit knowledge just as much as in verbal knowledge. And I saw my work as an attempt to articulate that learning, to explore and assess its impact, partly because it is such a strong methodological tool although relatively untested. As I practiced the research I realized that I was also researching the practice, and have slowly developed an understanding of the connection of early Greek rhetoric to the movement of wrestling in the gymnasium. Many figures of rhetoric interconnect with the physical dynamics of bodies that respond to one another, and I have no doubt that both Greek and Daoist philosophies share a common interest and inspiration in practice as research. This insight may or may not result in a written article or monograph, but will continue to inform my study of the rhetoric and ethics of movement.

The primary aim of the projects is to produce research into new strategies and structures for democratic humanism and a more participatory democracy. The strategies and structures being developed come from arts practices, just as did the highly sophisticated strategies and structures for the classical argumentative rhetoric at the heart of liberal democratic discourse.[13] What I am interested in is not subversive, but alternative rhetorics. Subversion feeds into changing hegemonic structures and is very effective. But alternative rhetorics usually focus on building the energy needed for any kind of change in the first place. This action research is not about Mouffe and Laclau's "articulation,"[14] getting desire into representative forms that can be recognized, however valuable that is. It is more about Nicole Brossard's idea of installation which focuses on the "fiction-making" needed to install oneself into a place where we are recognized and valued for what we are,[15] or Daphne Marlatt's idea of the energy that mobilizes that making,[16] which is mostly about presence. This is the primary reason that I distinguish between "articulation" and "articulating" at the head of this essay. Mouffe and Laclau locate articulation within "discourse," and seem unconcerned with any activity that occurs outside of hegemony. In contrast "articulating" is more concerned with the process of work that occurs before a work or a person emerges into the hegemonic field.

Re-search

My own artistic practice as research is modest, yet it has contributed to my theoretical understanding and engagement with research being done in the field. It's interesting how I feel the need to write both those rather diffident phrases, probably because I've been brought up to downplay any practice that I've done in the arts, and this discursive excursion has been partly an attempt to value work that isn't often accorded value. I have the distinct feeling that the university tolerates my artistic and creative research because I continue to also produce bone-tiringly appropriate academic scholarship. However, there are increasing numbers of people who are involved in thinking through these issues, and this is a potentially exciting time to be building programs that encourage practice as research, value the arts, and promote the individual and social transformations of which they are capable.

Notes

1. *Can a Man be a Woman?* (1994–8); *Trying not to be a tragic subject* (1995–7); *Cooking the Books* (1996–2001); *Bodies in Trouble* (1997); *Face-Work: Coming to the End of*

the Line (1999–2001); *The Face, the Mask and Classical Tragedy in the Household* (2003–5); *De-Scripting Performance* (1995–2007); *How Analytical Thought Stops You Thinking* (1994–2007).

2. Lorraine Code, *Ecological Thinking: The Politics of Epistemic Location* (Oxford: Oxford University Press, 2006), 11, 73.
3. Jane Mansbridge, "Using Power/Fighting Power: The Polity," in *Democracy and Difference: Contesting the Boundaries of the Political*, ed. Seyla Benhabib (Princeton, NJ: Princeton University Press, 1996).
4. Written version: Lynette Hunter, "Standpoint Theory Approaches to Recent Canadian Autobiographical Text," in *Autobiographies*, ed. M. Dvorak (Rennes: Presses Universitaires de Rennes, 1997).
5. Written version: ibid., "The Puppeteer. Being Wedded to the Text," *Open Letter* 9, nos 5–6 (1996).
6. Written version: ibid., "Learning to Read Writing from Non-Ruling Relations of Power," *Zeitschrift fur Kanada-Studien* 18, no. 1 (1998).
7. Simone Forti, *Handbook in Motion: An Account of an Ongoing Personal Discourse and Its Manifestations in Dance* (The Press of the Nova Scotia College of Art and Design and Contact Editions, 1974).
8. *See* T. Smalec, "Review (Re: Practice as Research in the U.S.)," ASTR-L Discussion Listserve. Currently offline.
9. Written version: Lynette Hunter and Susan Rudy, "Labour Notes," *Open Letter* 10, no. 8 (2000).
10. Written version: Lynette Hunter, "*Face-Work* and Going to the End of the Line with Frank Davey's Writing," in *Axial Writings: Transnational Literatures, Cultural Politics and State Policies*, ed. C. Batt, E. Boehmer, and J. MacLeod (Wollongong: Kunapipi, 2003).
11. Written version: ibid., "bp Nichol's Selected Organs: De-Scripting Performance," *Open Letter* April (2008).
12. Access accreditation is provided through the Open College Network in the UK.
13. I am increasingly interested in how Plato drew an understanding of rhetorical figures and schemes, if not strategies, from the gymnasium. On a related topic, *see* Deborah Hawkee, *Bodily Arts: Rhetoric and Athletics in Ancient Greece* (Austin: University of Texas Press, 2005).
14. I distinguish between "articulating" and "articulation" as used by Chantal Mouffe and Ernesto Laclau, in *Hegemony and Socialist Strategy: Towards a Radical Democratic Politics*, trans. W. Moore and P. Cammack (London: Verso, 1985), 113.
15. Lynette Hunter, "The *Inédit* in Writing by Nicole Brossard: Breathing the Skin of Language," in *Nicole Brossard: Essays on Her Works*, ed. L. Forsyth (Montreal: Guernica, 2005).
16. Ibid., "What Is an Honest Man and Can There Be an Honest Woman? The Poetics of Daphne Marlatt in Context of Global Economic Pressure," *Open Letter* 12, no. 8 (2006).

35
Dramaturgy: Conceptual Understanding and the Fickleness of Process

Jon D. Rossini

With the increasing interest in practice as research, dramaturgy, already understood as *the* research space within traditional theatre production, seems a logical place from which to generate an understanding of this activity.[1] To understand dramaturgy as a kind of practice as research – as opposed to dramaturgical practice as a kind of research – it is important to understand a sense of on-going process that is usually, but not always, highlighted in dramaturgical work, and to clarify the scope of dramaturgy, itself contested. One understanding of a dramaturg's role, often labeled a literary manager in the United States, is to help shape a season's offerings and assist an artistic director in thinking through the vision of the institution.[2] A second commonly invoked role is stewarding a new play through development. In new play development, the notion of working with a writer to shape a piece provides the clearest evidence of dramaturgical creative activity, but this contribution is often contested in relationship to questions of authorship (how much credit should a dramaturg's influence receive) and is troubled by a presumption, sometimes accurate, that the dramaturg serves the institution first rather than the playwright's vision. This latter concern emerges problematically in attempts (conscious or unconscious) to satisfy the conventionalized expectations of an institutional space's predetermined audience.

Perhaps the most recognized and least controversial role is the production dramaturg. Typically, the production dramaturg works with the director and creative team to help bring a production to fruition. According to Mary Luckhurst:

> Production dramaturgs are theatre practitioners who form working partnerships with directors and are generally textual specialists of some kind. They develop the "concept" for the performance with the director in the pre-production phase, are present in rehearsal, and may also work on publicity, programme material or publication of the text. Such dramaturgs are concerned overwhelmingly with textual issues and with articulating *processes* of production.[3]

This pragmatic definition is widely shared. While many forms of dramaturgy may in fact offer rich possibilities, I start with Luckhurst's traditional definition in order to point to a smaller but crucial intervention that can be widely practiced as a form of practice as research. Luckhurst's definition contains three phrases that need to be investigated: "pre-production phase," "textual issues," and "articulating *processes*."

Concern with defining the role and scope of the dramaturg in a particular context has a pragmatic rather than a philosophical function since dramaturgs are often perceived as dangerous for performing custodial roles, not merely in terms of the institution but in terms of the text as a protected site. These "textual issues," understood both as a focus on research on language and selection of text, often get transformed into an anxiety about the primacy and sanctity of a text, exemplified in culturally protectivist stances manifest in relationship to Shakespeare's canonical status. While the Shakespeare industry is unfairly criticized for this practice, a strain of Shakespearean studies that insists on historical re-creation is often placed at the center of this form of protectionism.[4] In this model of dramaturg as policeman, the assumption is that a text is something in need of preservation and that it is the intellectual's role to sustain this protection, as if actors, designers, and directors are necessarily dismissive of the text. While this practice does occur, a more useful discussion focuses on shifting the orientation of the dramaturg in relation to the production team and the process of creation.

Assuming that the relationship between director and dramaturg has been successfully negotiated, the "pre-production phase" involves selection of the text and translation (if necessary) as well as research that helps support and develop the conceptual framework for a production. While this is an effective and exciting form of dramaturgy – working with a director to deepen and strengthen her vision by providing means to think about historical and cultural contexts, including production history – this practice is essentially supplemental to the material manifestation of a particular director's artistic vision. Ironically, this can result in a form of idiosyncratic textual policing where the dramaturg is placed in the service of facilitating and deepening the representation of a pre-established concept.

To counter this more subtle form of policing one must shift the emphasis towards a sense of process. Although Luckhurst suggests the dramaturg's investment in the "articulation" of process, the necessary and subtle shift is based on the continuation of that process. Most understandings of dramaturgy as process discuss the role of documentation and production memory and assistance with the realization of a text in performance. All of these are essential elements of production dramaturgy in its different manifestations, and most dramaturgs already accept process as a fundamental constituent of dramaturgical practice. Dramaturgy as practice as research, however, emerges within the space of the rehearsal, not merely as a precursor and frame for this period. In part this is a shift between the realization of a vision and the asking of a question through the process of staging a work of drama; the more the directorial and design processes are explicitly involved in the exploration of a research question, the less the dramaturg needs to reimagine her role as processual. However, within a traditional model of practice that does not explicitly frame itself *as research*, unfortunately questions are often articulated within very concrete and practical frameworks of achievement and success, or if they are more abstract, remain only implicit in the process.

To make the shift to dramaturgical practice as research and not merely the practice of research, one could employ a number of contemporary dramaturgical practices.[5] Regardless, a reorientation must occur, one that goes not only beyond assumptions of protectivism, fidelity, and supplementarity, but that begins to articulate its sense of the processual as necessarily provisional, an understanding that *the doing of a project does not merely answer but also generates questions that are in need of answers*. While many dramaturgs would vehemently resist the notion of any kind of fixed text or interpretive practice, to genuinely transform dramaturgical practice into practice as research requires a transformation or confirmation that emerges in the doing – an incomplete practice that exposes the lack of production even as it points to the possibility of another. This potentiality, a shift in the typical thinking of performance as measured in relationship to an existing paradigmatic understanding of the behavior or drama, emerges from an increasingly symbiotic process in which performance opens up new forms of knowing or recasts the terms to which the production is implicitly imaginatively compared – it changes the very frames for understanding the work itself. Of course, like any research shift, a change in frame is only rarely a shift in paradigm, but is instead a subtle renegotiation of the cultural contexts under which this event can be understood to shift the philosophical and aesthetic questions that emerge.

To examine this process I will outline a potential moment of practice as research in both real and abstract terms to think about how this shift in thinking can happen. In November 2006, the University of California, Davis staged a production of William Shakespeare's *Cymbeline*, for which I served as dramaturg. For the most part, my function was a typical manifestation of a minimalized production dramaturgy: I did background critical research on the play, its historical production context, and the historical context it attends to, in order to supplement the reference materials of the critical edition used as a preliminary performance text; along with my assistant, I compiled a program supplement that provided some of this historical and contextual material for audience consumption. I also attended rehearsals and made suggestions to the director in addition to working individually with actors to help them understand the implications of their lines and their characters. And, eventually, I assisted the director in cutting the text prior to production.

Cymbeline, a less popular play, troubled both by its fantastic structure involving an internal pageant/masque and its relatively new status as a "romance" with a story taken from Holinshed's chronicles and a tragic label in the First Folio, is currently explored for its relationship to geography, history, and nationalism – an emerging sense of Britain in the face of Roman demands for tribute. Clearly a political allegory on many levels, the play offers a second and more important theme – the limits of perception and self-understanding. This theme has been widely remarked upon by scholars, most thoroughly and recently by Cynthia Lewis (1991). While questions of perception can be articulated as a banal theme of any Shakespearean work, what makes its presence in *Cymbeline* so crucially important is a recognition of limited self-understanding as a constitutive part of identity. Lewis argues:

> [t]hough they may appear to accumulate in haphazard and nondescript fashion, Shakespeare in fact plots instances of misperception along a hierarchy of difficulty: not all perceptual puzzles are so easily solved as others. This design contributes significantly to Shakespeare's characterization in that the peculiar type of misperception to which a character falls prey reveals much about that character's intelligence and psychological or emotional state.[6]

This issue becomes most clearly manifest in the beautifully condensed and unique use of the word *imperceiverant* (or *imperseverant*), a term that occurs only in IV.i and has historically meant to different interpreters "incapable of perception" or "incapable of perseverance." Ironically spoken by Cloten, perhaps the least self-aware and most abusively derided character in the play, the adjective is used to describe Innogen, the central protagonist whose refusal to submit to Cloten's attention has secured his wrath. For Lewis, the various forms of misperception occurring in the work (interestingly, she does not specifically attend to Cloten's frustrated articulation) point towards a concluding investment in faith beyond the bounds of human perception – "an arrogant belief in the self has been transformed into unqualified, humble faith reposed in others."[7]

However, rather than reducing "imperceiverant" to a sign that points to the need for a different mode of perception, combining the possibility of a lack of perception tied to a lack of willingness to persevere allows for a question to emerge in relationship to the nature of character within the play itself. This combination of fickleness and a failure of perception emerged through the Davis rehearsal process as a key element for "understanding" the limited self-explication that characters can manifest, the problematic historical narratives, and the improbable and unbelievable events characteristic of the romance genre. This combination, in which perception and perseverance are potentially connected, suggests an understanding of character development and identity manifest in process, but also limited by the very terms of the process through which it manifests.

In the case of the production at UC Davis, the understanding of a particular way of thinking about cognition and perception as central elements of *Cymbeline* emerged not merely from a nuanced reading of the text, but observation of the rehearsal process and the sustained confusion of student actors in relations to the goals and motivations of their characters. This was compounded by a difficulty reconciling inconsistent emotional development and rapid transformations in attitude and point of view. Although I was aware of the theoretical thematic of questioning of perception and identity in this play prior to the rehearsal process, I did not realize the extent to which this conundrum permeated the very enactment of the play itself. Ironically, we found that as a means of negotiating the motivations of their characters, student actors were best served by acknowledging the limits of their knowledge and facing the possibility that not only do the characters not fully understand their motivations, but that the actor's themselves need to retain a fundamental inability to perceive in order to understand the position of their character. Rather than adopting a position where the actor has more understanding than the character, the actor must accept the limited and confused knowledge of the character as the contours for their understanding, not attempting to impose an artificial coherence upon them.

What exactly is at stake here in the embodied recognition of misunderstanding? Examples of misconception abound, from Cymbeline's inability to recognize the machinations of his queen and the relative value of his daughter's suitors, to the inability of his sons, Cadwal and Polydore, to recognize their own noble birth despite, as is voiced by their adoptive father and abductor, the fact that their nobility constantly shines through (III.iii). The play's key figure is not the historical king of titular fame but rather his daughter, Innogen, who is busily fending off the advances of her undesirable though well-placed step-brother Cloten and pining for her exiled love, Posthumus.

In his longest soliloquy, Cloten, a man schooled in self-ignorance, is alone in the woods seeking to kill Posthumus, defile and dominate Innogen, and destroy Posthumus's clothing in front of her, an odd gesture that emerges from the specific shame Cloten feels for an earlier insult from Innogen suggesting that Cloten is less important than Posthumus's "meanest garment" (II.iii.127). Attending to his speech is crucial, because the phrase he invokes is in the context of declaiming his own superior attributes, established physical attractiveness, cultural and political position, as well as his lack of understanding how and why Innogen could remain ignorant of them. He rails, "yet this imperceiverant thing loves him in my despight" (IV.i.11–12). The use of "despight" also creates an interesting verb structure, suggesting the shifting agency that accompanies her choice not to love him – her lack of perception is also a lack of persistence and a perceived moral failure from Cloten's point of view. Of course, the space where this concept is voiced is itself an ironic and problematic one. Soon to die at the hands of Guiderius, Cloten's voicing of the "not perceptive" suggests both a complaint and a condition. For the actor playing Cloten, it was necessary that he be incapable of perception, not as a purely farcical articulation of character, but as a fundamental condition.

This is an entire world where the question of perception is a condition of partial knowledge that must be embraced. The anxiety about national emergence and self-determination is tied up within this reality as are the extended dream-like sequences of the play, both the extended masque and Iachimo's visit to Innogen's bedchamber. The difficulty of negotiating the plot in production, especially what is often viewed as an unnecessarily burdensome final scene, is symptomatic of the problematic status of knowledge. The conclusion of the play is difficult to stage in part because of the continuous revelations that emerge only to be followed by an odd supplementary interpretation of the entire play articulated by a soothsayer. To generate the genuine transformation, characters must remain in a fundamental state of misunderstanding about their own condition even as actors are informed. In this moment where the actors are placed in a situation of greater knowledge than their characters, it becomes necessary to enact a world in which one's own perception requires a form of authorization for validation and acceptance – even the revelations of first-hand witnesses must be supported by a power greater than the self.

Importantly, this process of voluntary submission echoes the politics of the play itself as Cymbeline voluntarily submits to the demands of Rome, and thus the pragmatic choices of actors emerge not just from a reading of a given line, but a broad understanding of the conceptual framework of the play itself. In the space of practice as research, it is not merely the content of a character that emerges from the rehearsal process, but potentially the very tools through which an actor engages the play. Embracing *imperceiverance* became a way of thinking on and through the character that allowed some actors to balance and inhabit the fundamental contradictions and apparent confusions of their characters, and finding ways to limit the truth of their perception provided a means of sustaining the revelatory energy of the performance. In thinking on and through the word, *imperceiverant*, as a spectator and participant in a development process, the negative intersection of perception and perseverance offers a new way of ordering and holding together not only the seemingly disparate elements of the play, but also suggests a new pedagogical project for the work for a contemporary audience and those participating – a rethinking of one's mode of understanding as shaped inexorably by the accepted gift of another's interpretation. By thinking about dramaturgy not merely as a supplement, but as a process that is never completed within the space of production, the possibility of dramaturgy within a practice as research as opposed to a practice of research model becomes possible.

Notes

1. For good overviews of dramaturgical practice and attitudes towards research, *see* Judith Rudakoff and Lynn M. Thomson, eds, *Between the Lines: The Process of Dramaturgy* (Toronto: Playwrights Canada Press, 2002); Susan Jonas, Geoff Proehl, and Michael Lupu, eds, *Dramaturgy in American Theater: A Source Book* (Fort Worth, TX: Harcourt Brace College, 1997); Cathy Turner and Synne K. Behrndt, *Dramaturgy in Performance* (New York: Palgrave Macmillan, 2008).
2. In Europe, especially Germany, the dramaturg in this role would be responsible for the institutional artistic vision.
3. Mary Luckhurst, *Dramaturgy: A Revolution in Theatre* (Cambridge: Cambridge University Press, 2006), 205.
4. Andrew James Hartley, *The Shakespearean Dramaturg: A Theoretical and Practical Guide* (New York: Palgrave Macmillan, 2005).
5. *See* D.J. Hopkins, "Research, Counter-Text, Performance: Reconsidering the (Textual) Authority of the Dramaturg," *Theatre Topics* 13, no. 1 (2003). *See also* the chapter "Millennial Dramaturgies" in Turner and Behrndt (2008). The notion of processual and interactive dramaturgy as an element of performance is also articulated in Eugenio Barba, "The Deep Order Called Turbulence: The Three Faces of Dramaturgy," *TDR* 44, no. 4 (2000).
6. Cynthia Lewis, "'With Simular Proof Enough': Modes of Misperception in *Cymbeline*," *Studies in English Literature, 1500–1900* 31, no. 2 (1991): 345.
7. Ibid.: 360.

36

Collisions in Time: Twenty-First-Century Actors Explore Delsarte on the Holodeck[1]

Sharon Marie Carnicke

Tremendous potential resides within practice-based methodologies for researching the history of acting. It bears repeating, however obvious it may seem, that acting, as a discrete artform, has been particularly resistant to documentation and analysis. Nicol Williamson has compared the "evanescence of acting" in live theatre to "rain falling into a river; it's beautiful and it's gone."[2]

In light of acting's "evanescence," traditional scholarship has unsurprisingly tended to circumvent performance. Historians initially found objective analysis easier for those artforms that surround the actor and leave behind more tangible evidence: such as playscripts, the architecture of theatres, design drawings, and so on. Scholars who do examine the theories and mechanisms of acting struggle to find vocabularies that avoid the often subjective language of journalistic reviewers – especially as subjectivity in analysis tends to draw reproach from external referees whose job it is to vet the rigor of scholarship.

In the late twentieth century, scholars who sought to get beyond these difficulties opened a new window onto the history of acting. Foundational work in this direction was done by Raymond Williams, who saw that plays encode performance, and Richard Schechner, who reframed performance as behavior subject to study in the same ways that anthropologists study other forms of human behavior.[3]

In my view, practice-based research into the history of acting represents the next methodological step in this continuing investigation. But, as this anthology demonstrates, "Practice as research (PAR) and practice-based research (PBR) – and 'research through practice,' 'research by practice,' 'performance as research' – are contested terms that resist close definition."[4] A dizzying array of phrases and acronyms brands this young field – a field that means different things to different people, even as it consistently bridges the world of ideas to the material world of performance.

For me, practice-based research means bringing traditional scholarship together with the practice of acting in order to investigate its "evanescence."

Such collaborative work can best address the elusive questions: How can scholars and theoreticians better speak about practice? How can theatrical history and acting theories better illuminate the world for practitioners? How can theory prompt practice? How do practitioners generate theory? As a scholar who has acted and danced, I feel in my bones how embodied, tacit knowledge can enhance what scholars learn through traditional, archival research. Conversely, I see value for performers in scholarly encounters with the written past.

In order to explore some of these questions and possibilities, this article examines the efficacy of practice-based methodology in a brief case study of an ongoing project that brings the work of the nineteenth-century acting theorist François Delsarte into the postmodern world of virtual reality.

Delsarte on the holodeck

On the television show *Star Trek: The Next Generation,* the twenty-fourth-century crew of the starship *Enterprise* has a *holodeck* at their disposal. In the science of this fiction, a computer driven "theatre" projects holographic imagery through a matter conversion subsystem that creates credible three-dimensional, interactive environments.[5] A crewmember can choose a program, enter the *holodeck,* and interact with a full cast of virtual characters who complete the immersive illusion.

What if engineers could actually build such a holodeck? And what if acting teachers had to train its intelligent virtual agents to act? At the Institute for Creative Technologies (ICT) at the University of Southern California (USC), computer scientists of the "Mission Rehearsal Exercise Project" have, in fact, built a "prototype of a holodeck-like environment."[6] The overall goal is to create a tool whereby a person can "rehearse" the skills of interaction and negotiation. The project's director envisions a "learning system where the participants are immersed in an environment where they encounter the sights, sounds, and circumstances of real-world scenarios. Virtual humans act as characters and coaches in an interactive story with pedagogical goals."[7] The 2001 ICT *holodeck,* already outdated by the forces of history, brings a Bosnian village to life in an effort to figure out how such a "learning system" might train military personnel faced with peacekeeping efforts in another culture. The ICT room is decked out with screens that encircle the walls, ride technology to simulate movement, and speech recognition software to allow interaction with virtual characters (see Figure 8).

In 2003, I joined the team as theatre consultant to assist with another project goal: "to create general computational models of the interplay between affect, cognition and behavior" so that "the behavior [of a virtual character] also suggests an underlying emotional current."[8] In other words, I was asked to help "teach" virtual actors to act with emotional credibility. My research partners in this endeavor became Jonathan Gratch and

Figure 8 The "holodeck" at the Institute for Creative Technologies (ICT) at the University of Southern California. Photograph by William Swartout. Reproduced by permission of the project director, William Swartout

Stacy Marsella, specialists in computer science, and Albert Skip Rizzo III, a psychologist and leader in virtual reality.

I admit that I joined the team with some ambivalence. On the one hand, I worried that this project might ultimately replace human actors. Quickly, the immense complexity of our work reassured me that virtual actors would not take the place of humans any time soon. On the other hand, I hoped to learn much about technologies, like motion capture, that already effect the conditions under which professional actors work.[9] Unexpectedly, I found a rich interdisciplinary environment open to practice-based research methodologies. Perhaps this is so because engineering and acting share a basic pragmatic mindset.

We began by asking ourselves a single question. How is emotion performed and expressed through the human body? Psychologist Paul Ekman has determined that certain facial gestures consistently express key emotions across gender and cultures.[10] Could we do with the body what Ekman had done with the face? If we could find bodily gestures that express emotion as consistently as those found by Ekman in the face, then computer scientists might be able to create software to direct virtual actors to use these gestures appropriately within the *holodeck* scenario. In other words, they could write programs that would allow virtual actors to interact emotionally with humans.

The process of our work has just begun, but already suggests the value of the actor's tacit knowledge in scientific inquiry. Like other similar collaborations across the sciences and arts, our project creates a space where "[i]nsights from the arts, including narrative theory, animation, dance, theater, and film" are as important as "research in artificial intelligence, psychology, human ethology, and linguistics to name a few fields."[11]

At the outset, we decided on two avenues of investigation. The first was to hire four actors to assist us: two men, Keith Day and Setrak Bronzian, and two women, Jennifer Kaplan-Dreyfus and Mitch Jacobson. For two months in 2003, they performed a series of improvisations, which were videotaped for analysis.

The second decision involved a radical idea to go "back to the future." No doubt readers of this book can easily understand how unhelpful a Stanislavsky-based approach would be for virtual actors, who have no psyches to activate through techniques like the "magic if." As Ed Hooks writes, "Actors create emotion – largely internally – in the present moment, while animators describe internal emotion through the external movement of their characters."[12] Like the animator, the ICT project needs to separate physical gestures into their multiple components (spatial shapes, temporal rhythms, force, and direction). While contemporary actors ride the dynamics of time like a roller-coaster in order to create emotionally expressive behavior, the ICT group must stop time and dwell in the moment long enough to capture and describe behavior. Moreover, as a scientific enterprise, the group also needs to collect measurable data about the encoding and decoding of emotion through the body. To negotiate between these disparate modes of working, I suggested that we (re)turn to the principles of François Delsarte (1811–1871), who worked on the physical aspects of acting.

In an effort to improve training, Delsarte diligently studied the relationships between physical gesture, emotion, and language. Over time, he became the most significant acting teacher in Europe, and until Stanislavsky brought the Moscow Art Theatre to the United States in 1923, Delsarte's methods were the predominant form of American actor training.[13]

At base, Delsarte sees movement as a "semiotics" – a sign system that can be "read" by observers. He also recognized that physical "signs" come from various sources: some gestures express individuality; some are socially determined conventions, like waving "hello"; and some may be biologically connected to our emotional reactions.[14] In these broad strokes, he seemed to anticipate Ekman's findings about facial gesture.

Delsarte developed a maddeningly complex, yet surprisingly systematic vocabulary for describing movement and for predicting how gestures might be read as emotional expression. Consider how Delsarte analyzes the movement of hands in space: visualize an imaginary cube in front of you, and grasp it from each possible surface – use two hands to contain the outer surfaces; push a hand outward against its inner surface; bring a hand upward to its lower surface; explore every possible way of grasping, pushing against, and containing this cube. Delsarte studies each of these movements and predicts its emotional expressivity.[15] Such logical multiplication of possibilities proved more helpful to ICT than other physically based actor-training programs (such as that of Rudolph Laban, which has been used by other scientists in experiments on gesture). As theatre scholar Patsy Ann Hecht observes, physical training systems after Delsarte became progressively less complex.[16] While actors find these later approaches more manageable, for us they proved too limiting.

After unsuccessful attempts to use Delsarte to prescribe emotionally expressive movement, we realized that we could not expect contemporary actors to master Delsartean complexities for the sake of the experiment. Instead, the *holodeck* team decided to use Delsarte as the lens for our analysis: the four actors improvised short scenarios designed to elicit emotional responses, then we studied the videotapes with Delsarte's principles in mind.

During the improvisations, I became a kind of interdisciplinary translator. The scientists spoke to me in terms of Delsarte and data measurement; I relayed this information to the actors using a more familiar language of "given circumstances," "objectives," and "actions." This approach worked well.

In studying the tapes, we discovered that Delsarte's insights into emotive gestures were everywhere present in the actors' work. By March, we had begun to focus on the basic patterns and sequences that Delsarte calls "Orders of Movement." In brief:

- **Oppositions** occur when any two parts of the body move in opposite directions simultaneously. Delsarte sees this type of movement as expressive of force, strength, and either physical or emotional power.[17]
- **Parallelisms** occur when any two parts of the body move in the same direction simultaneously. Delsarte suggests this type of movement can express physical weakness or intentional planning.[18]
- **Successions and Reverse Successions** are defined as "any movement passing through the body which moves each part it moves through (in a fluid wave-like motion)." *True successions*, which Delsarte identifies as expressive of sincerity, travel from the face into the rest of the body. In contrast, a *reverse succession* flows backward through the body, and for Delsarte, signifies falsehood.[19]

From the videotapes, we selected a series of two- to three-second clips that featured a clearly executed gesture. The clips were edited to show only one person moving against a blank wall and the soundtrack was removed, eliminating as much contextual information as possible. Our plan was to test the emotional legibility of isolated physical gestures by asking viewers what they saw. We showed the clips to 23 students at USC during the spring term of 2004 with tantalizing results. The responses were remarkably consistent with each other with regard to emotional content, general context, and the type of interpersonal relationships created between the visible actor and off-screen partner.

For example, after watching Setrak Bronzian perform an opposition through the body with one arm brought forward, four viewers thought they saw "the man yelling at his child" despite the lack of sound, seven thought he was either "telling another person what to do" or "trying to make a point," and five viewers considered him to be "a professor" or "teacher" giving a lecture. All agreed that he held the position of power within the relationship. After viewing another clip of Mitch Jacobson with feet parallel and hands clasped in front of her chest, most viewers assumed she was "talking to a boy she likes," "waiting" either for "an answer" or "for a surprise which she is looking forward to," and that she seemed "eager" as if she were "anticipating something like good news." Seven responses characterized her as "excited" and another three as "surprised." Most viewers saw her emotional state as subject to her invisible partner.

Equally remarkable to me was the differential reception with and without sound. Our team saw and heard actors who were natural in their behavior; the viewers of the silent clips generally felt that the actors were overacting. This result prompted me to look with new eyes at the silent films of the 1920s, whose actors had been largely trained via Delsarte.

The sample of viewers was too small for scientific publication and so the data was not measured. Nonetheless, this preliminary study was enough to confirm the value of continuing our research with actors. For the purpose of this volume, it also suggests how actors' tacit knowledge can participate productively in interdisciplinary research of this sort.

Some conclusions

As "Delsarte on the Holodeck" suggests, crossing distant disciplinary boundaries with practice-based research not only makes methodological sense, it also brings actors productively into scientific research.

Within the discipline of theatre, however, practice-based methodologies cross a more resilient border between theory and practice. The field of theatre arts has long been divided into scholars and practitioners who have rarely collaborated. The gap that exists between scholars and actors sometimes translates into a gulf between how scholars perceive acting and how actors create performance. For example, scholars who study the feminist theatre group, Split Britches, tend to value the company's work for its Brechtian estrangement – for the actors' ability to comment upon rather than merge with their characters. As Elin Diamond writes, Split Britches "enables [the spectator] to see a sign system *as* a sign system."[20] However, Deb Margolin, the company's founder, stresses their need to identify with the characters, a desire that flies in the face of Brechtian theory: "We cannot perform what we do not imagine, and we cannot imagine that which is not within us to be conceived."[21] This points to a significant gap between reception and creation.

The gulf between reception and creation is constructed by the different processes that undergird theory and practice. In his work on writing pedagogies, Stephen M. North notes that practice depends upon the accrual of information, which creates a kind of toolbox filled with useful, if sometimes contradictory tips. In contrast, theory depends upon throwing out hypotheses once they are proven to be wrong. North envisions these two modes in architectural terms: practice is like a Victorian house "wing branching off from wing, addition tacked to addition, in all sorts of materials,"[22] whereas theory is like the elegant, sleek lines of modern architecture.

For theatre art, bringing these two modes of thought together in practice-based research is highly productive. On the one hand, books are an imperfect way to understand an art based largely on tacit knowledge. On the other hand, practice without reference to history proves equally imperfect in transmitting acting theories to future generations of practitioners. Practice-based research can offer actors valuable new tools that would otherwise remain forgotten. I believe that practice-based methodologies can enrich our knowledge of theatre even as it makes possible interdisciplinary and applied research. I envision a twenty-first century full of new knowledge generated by work that crosses boundaries between discrete disciplines and between theory and practice.

Notes

1. For their assistance in preparing this chapter, I want to thank Jonathan Gratch and Stacy Marsella at USC; Shannon Rose Riley, who suggested I write it; and Mary Joan Negro for her comments.
2. Nicol Williamson, *Nicol Williamson: Critics and the British Star Discuss His Controversial Hamlet* (North Hollywood: 1969), Sound Recording.
3. Raymond Williams, *Drama in Performance* (London: Frederick Muller, 1954); Richard Schechner, *Between Theater and Anthropology* (Philadelphia: University of Pennsylvania Press, 1985).
4. Baz Kershaw, "What Is Practice as Research?," Arts and Humanities Research Board, University of Bristol, *see* http://www.bristol.ac.uk/parip/faq.htm#menu.
5. Rick Sternback and Michael Okuda, *Star Trek: The Next Generation, Technical Manual* (New York: Pocket Books, 1991).
6. W. Swartout et al., "Toward the Holodeck: Integrating Graphics, Sound, Character and Story," in *Fifth International Conference on Autonomous Agents* (2001), 409.
7. Ibid.
8. Stacy Marsella and Jonathan Gratch, "Modeling the Interplay of Emotions and Plans in Multi-Agent Simulations," in *Proceedings of the 23rd Annual Conference of the Cognitive Science Society* (Edinburgh, Scotland 2001), [1].

9. For a detailed picture of how technology is changing the ways actors work, *see* Andy Serkis, *The Lord of the Rings Gollum: How We Made Movie Magic* (New York: Houghton Mifflin, 2003).
10. Paul Ekman, *Emotions Revealed: Recognizing Faces and Feeling to Improve Communication and Emotional Life*, 2nd edn (New York: Henry Holt, 2003), 3.
11. Stacy C. Marsella et al., "An Exploration of Delsarte's Structural Acting System" (paper presented at the Proceedings of the Sixth Annual International Conference on Intelligent Virtual Agents, Springer-Verlag Berlin Heidelberg, 2006), 91–2.
12. Ed Hooks, *Acting for Animators* (Portsmouth, NH: Heinemann, 2000), 2.
13. One of Delsarte's students, Steele MacKaye, founded the first professional US acting school in 1884. *See* James H. McTeague, *Before Stanislavsky* (Metuchen NJ: Scarecrow Press, 1993), 1–43.
14. François Delsarte, ed. *Passions Signs: Signs of Passion*, Delsarte System of Oratory (1893 [n.p.]), 459–62.
15. Ted Shawn, *Every Little Movement* (New York: Dance Horizons, 1963), 42–3.
16. Patsy Ann Hecht, "Kinetic Techniques for the Actor" (Dissertation, Wayne State University, 1971), 264–5.
17. Shawn, *Every Little Movement*, 33–4.
18. Ibid., 34.
19. Ibid., 34–5.
20. Elin Diamond, "Brechtian Theory/Feminist Theory: Toward a Gestic Feminist Criticism," *TDR* 32, no. 1 (1988): 85.
21. Deb. Margolin, "Mining My Own Business: Paths between Text and Self," in *Method Acting Reconsidered: Theory, Practice, Future*, ed. David Krasner (New York: St. Martin's, 2000), 132.
22. Stephen M. North, *The Making of Knowledge in Composition: Portrait of an Emerging Field* (Upper Montclair, NJ: Boynton/Cook Publishers, 1987), 27.

37

Living on the Edge: Alternate Controllers and the Obstinate Interface

Joseph Butch Rovan

What can one say about the apparently endless supply of new electronic instruments?[1] Born in labs or garages, they bear witness to a desire for new forms of musical expression. They are also a specific manifestation of performance as a catalyst for research. Scholarly conferences devoted to computer music research regularly feature papers by inventors of alternate controllers.[2] Some of these instruments are designed to harness new kinds of gesture; others extend existing instruments. If few make it to the stage, most never make it off the bench. One could almost see the past several decades of research and development as a long instrumental highway littered with the remains of discarded demos. In some cases, the technology no longer supports the concept, as new computer operating systems render the instrument unusable, or sensors break and cannot be replaced. In others, the instruments themselves have never become fully operative. What they have lacked, it seems, is a meaningful performance practice.

This is not to say that contemporary research into new instruments has entirely disregarded the performer. There is a large body of research that has focused on the gestures of performance as the basis for new alternate controllers. This includes research into the theory of gesture, sensor design, data acquisition, and mapping.[3] The dominant voices are scientists whose discourse focuses on technology in general and draws on the field of Human Computer Interaction (HCI) in particular. They have applied efficiency tests to new controllers, for example, measuring their value in the same manner one measures the usability of a computer mouse. Even when such researchers are performers, the published results tend to privilege technology over performance.[4] In this chapter, I argue for a different approach, one that combines research in gesture and technology with the knowledge gained from live performance. Such a performer-centric approach ultimately requires a change in the concept of what an alternate instrument can be.

Instruments, instrumentality, and obstinacy

Instruments are not always cooperative. Despite years of practicing, performers still have to deal with the unexpected. Sometimes the unexpected becomes an aspect of a known performance practice, like the various honks and unintentional multiphonics that are a central part of improvising on the saxophone: if the instrument seems to resist the player's best efforts, over time players have learned to accommodate these behaviors, developing ways to minimize, or even maximize, their effects. David Burroughs has commented on this phenomenon, describing the intriguing "unmusicality" of most instruments.[5] Aden Evens has argued that such resistance is vital to any instrument's creative energy: "The instrument presents a leading edge of indeterminacy," he writes.[6] The resistance of the interface "holds within it its own creative potential. [...] The interface must push back, make itself felt, get in the way, provoke or problematize the experience of the user."[7]

Resistance means effort, and Evens suggests that all performance technique aims to "feel and work in the resistance offered by the instrument."[8] In a similar way, Joel Ryan considers such effort an elemental aspect of all performance, the locus of "energy and desire, of attraction and repulsion" that defines "the movement of music."[9] From this, one could argue that resistance is inherent in the very idea of an instrument. And yet we know that many alternate controllers are created to minimize effort, either by focusing on gestural efficiency or by repurposing existing commercial devices like joysticks or writing tablets, which were designed, after all, to be as efficient and ergonomic as possible.[10]

In order for new musical interfaces to serve as a proper vehicle for musical performance, I would argue, they must recover some of the resistant quality that has been engineered out, and in this chapter I describe at least one attempt from my own work in which I tried to do just that. As I will show, such resistance needs to be felt not just at the level of the interface, but in the entire instrumental system. The "controller" cannot be merely a passive receptor of a performer's input, but must define a center of activity all its own. To be truly playable, in other words, the instrumental system must exhibit a certain degree of what I like to call obstinacy. When players find themselves interacting with an obstinate system, they begin to experience the productive effort that, as Ryan has argued, is elemental to musical creation. True performance occurs in this space – at the point where the performer is forced to live on that edge, negotiating the "play" between body and instrument in the movement of music.

I have been dedicated to living on this edge in my own creative work. In my instrument designs I have been especially interested in the capacity of the instrumental system to exhibit obstinate behaviors, acting in independent and at times indeterminate ways. The most creative systems, in my view, have the ability to become self-reliant, resisting the gestures of the performer to do their own thing. They may include a means by which the performer can regain control, but it is in such moments of struggle that the performer/instrument relationship becomes most meaningful: when the performer is no longer the controller but also controlled.

If resistance is thus defined by the behavior of the whole system, our very concept of the "alternate" controller must evolve to include this larger arena. All parts of the instrumental system point, then, to an embodied potential for new and emergent behaviors; all parts are, in effect, performative. I like to think of this comprehensive controller – with its built-in obstinacy, its sensors, its software, and its sounds – as a macrocontroller. As a new, more comprehensive electronic instrument, the macrocontroller thus requires us to conceive a more comprehensive performance practice.

Performing the macrocontroller

Playing any instrument requires learning the basic gestures that make it work, and that, in turn, requires differentiating between movements seen to be expressive, and others that are more utilitarian. Expressive gestures are by definition tied to sound production; utilitarian movements (or what I call "maintenance" gestures) maintain or change the instrumental state.[11] In the case of new electronic instruments, examples of such maintenance gestures might include the pressing of a foot pedal to change instrumental configurations, or the movement of a hand to zero out the signals coming from a sensor. Expressive gestures, on the other hand, are those that shape the musical performance; they are read as performative. They may range from small finger movements to a full body gesture; or they might be more felt than seen, as in an intake of breath or the tensing of a muscle.

This duality between the expressive and the utilitarian naturally creates a bifurcation of the performing body, where one part of the body (hands or arms or torso) is seen to create sound and the other is seen to perform menial labor. The condition raises the more basic question of gestural legibility. Should there always be a perceivable relationship between the performer's gestures and the resulting sound? If we look to acoustic instruments as a guide, we could argue that both expressive and maintenance gestures are necessary: in other words, audiences and performers expect that there will be moments where the relationship between movement and sound is clearly visible, along with those where it is not. Expressive gestures serve to orient an audience, highlighting structural aspects of the music or of the instrument itself. They can also be exaggerated for dramatic effect.[12] A performance comprising only expressive gestures is logically impossible and yet, with new musical interfaces, audiences often expect the performance to be completely transparent. Perhaps for this reason some laptop performers have gone in the opposite direction, aiming for opacity – attempting to downplay the very idea of "performance." In such cases the performer becomes an operator of the technology and the issue of gestural legibility becomes moot.

In my own design and use of alternate controllers, I see the performer's relationship to the audience as a crucial aspect of what makes the controller expressive. In working out this relationship, it has become clear how central performance is to my own research and development. In the remainder of this chapter I briefly describe two different instrumental systems developed over a ten-year period, which work through this relationship with varying results. In the first, the controller required me to develop a more elaborate performance practice in order to make the instrument's expressive gestures more immediately readable to the audience. This practice, once developed, ultimately revealed a basic limitation of the system. I follow this example with a more recent instrumental system that offers an alternative solution.

Alternate routes: COLLIDE and MiMICS

In 1997, while a researcher at IRCAM, I designed a custom glove controller system that I used to create an interactive piece based on a text by Archie Ammons called "Continuities."[13] I later extended the glove system for use in a more elaborate work involving interactive music, movement, and video, which I called COLLIDE (2002).[14] I continued to revise this piece over a two-year period in an attempt to make the relationship between instrument, image, and gesture more clearly understandable to audiences, and to myself, in the act of performance.

The controller involves two gloves: one for the right hand with several force-sensitive resistors (FSRs) that react to pressure and an accelerometer that measures acceleration and rotation; and another for the left hand with a reflective strip that interacts with an infrared proximity controller. Whereas the right-hand gestures are free, the left-hand gestures are relative to a fixed position. This physical tension becomes a focal point for the audience during performance, as they follow my attempts to negotiate the opposition.

When I created COLLIDE, I imagined a difficult piece, and an instrumental system that would require constant physical activity to keep it sounding. The meaning of the piece thus resides in the tension between the sounds, images, and gestures. The piece is composed in several "scenes," in which algorithms governing the relationship between all three change.[15] Some of these relationships were meant to be clearly understandable to the audience, while others were chosen to be obscure; I worked with a choreographer in order to refine the way I projected all these possibilities. Eventually, I adopted an illuminated ball as a kind of prop for use in performance. This not only provided a surface on which to "play" the FSRs with my right hand, it also created a weight that gave momentum to the formerly empty-handed gestures, while serving as a visual complement to the images playing across the video screen. Nevertheless, while the illuminated ball did increase the legibility of certain gestures, as a prop it had limited utility: it could not affect the performance of the controller itself. And if, as I have said, the most creative interactive systems have the ability to become self-reliant, exhibiting their own behaviors – their own resistance – then ultimately the glove system fell short of that goal. The glove was expressive but, even with the ball, it could not produce that quality of resistance that made me, as a performer, interact with it as a true "instrument." My next project attempted to build on what I had learned from the glove system in order to capture this missing dimension.

The result was a comprehensive controller that I call the Macro-Instrument Musical Interface Confrontation System (MiMICS). In one respect, MiMICS can be described as an extended instrument, expanding the capacity of an existing acoustic instrument via added hardware and software (in the current version it is based on my alto clarinet).[16] One of my goals, however, was to create a macro-instrument that allowed the performer to exploit normal playing techniques while also developing entirely new ways of playing on a familiar instrument. The MiMICS system thus combines aspects of my glove controller with an extended-instrument paradigm that relies on a clear vocabulary of alto clarinet gestures. In other words, it responds as I interact with my alto clarinet in a conventional way, while it also allows me to use the horn as an alternate controller, through non-conventional, full-body movements similar to those I used in COLLIDE. Moreover, as a system it is adaptable to any acoustic instrument.

In technical terms, the MiMICS system comprises several hardware and software components: a custom-built sensor interface with wireless transmission; multiple sensors, a contact microphone, and a wireless audio transmitter mounted directly on the instrument body; and a custom software environment that controls data mapping and audio processing.[17] The system also provides what is known as "haptic" feedback; that is, the performer experiences tactile sensations that offer an impression of the kind of information being captured by the sensors at any given moment in the performance.[18]

One of the advantages of the design is that I am able to move fluidly between different software configurations of the instrument, simply by using the normal performance gestures associated with the alto clarinet. I can also reimagine my horn in a different way: its long body has properties that I can exploit through extreme gestures that activate other audio processing routines. Here, the sensor interface allows me to control the real-time audio with a completely different set of expressive gestures. The alternation between traditional and non-traditional use of the alto clarinet amplifies the aspect of suspense – even surprise – that audiences expect from all meaningful performance (see Figure 9).

MiMICS can also create suspense and surprise for the performer, and this is another crucial aspect of its expressivity. I deliberately created the system to behave in unexpected ways, promoting what might be called

Figure 9 Joseph Butch Rovan in performance with MiMICS, NIME 2007, New York. Photograph by Mark Cetilia. Reproduced by permission of the photographer

"situational resistance." The controller has the capacity to interrupt the virtual connections within its own data-mapping engine, so that it will cease to respond to my gesture, and instead begin to generate gestural data on its own. This condition thus replicates at the level of the system the kind of resistance that, as I discuss above, defines a basic property of all instruments, a property that requires the performer to confront, and engage with, the instrument in a distinctively physical way. In performance with MiMICS, I have the ability to regain control of the system at these moments – in effect, to reconnect the virtual connections – by resorting to a class of extreme gestures: tilting, or spinning, or thrusting the horn through space. This makes the act of playing the controller far more dramatic, and requires that I continue to develop my technique on the horn in order to make the most of every performance situation. The MiMICS system approaches the condition of a true instrument not only in the way that it mimics the resistance necessary to all instrumental performance but also, more importantly, in the way that it foregrounds performers themselves in its revised approach to hardware and software design. Here, the performer is the impetus for new research. And, because the instrument has been designed with the performer in mind, it creates the conditions for its own survival, through a performance practice that the instrument itself, in a sense, demands.

Conclusion

Historical performance practice emerged in the late twentieth century as a field of research that hoped to ensure the viability of certain repertories from the past. Modern musicians sought to learn more about old instruments and how they were played, in order to give historical repertories a second life in a new century. We need to gather the same kind of evidence about electronic instruments today if we expect them to have a life in the twenty-first century and beyond. And, while we can rest assured that researchers will continue to develop new demos and publish technological accounts in conference papers, we will not have the full story about these alternative interfaces and their music until we hear much more from the performers who have actually used them.

Notes

1. I dedicate this chapter to the memory of Michel Waisvisz (1949–2008), an inspiring musician who lived on the edge. He enlivened the field of experimental music with his ever-innovative instrument designs but taught us even more through playing them.
2. "Alternate controller" refers to an electronic performance interface other than a MIDI keyboard that enables the production of sound. Examples include virtual wind instruments, electronic drum pads, data gloves, joysticks, writing tablets, and other custom interfaces; such conferences include the International Conference on New Interfaces for Musical Expression (NIME), the International Computer Music Conference (ICMC), and the National Conference of the Society for Electro-Acoustic Music in the United States (SEAMUS).
3. *See* Eduardo R. Miranda and Marcelo M. Wanderley, *New Digital Musical Instruments: Control and Interaction Beyond the Keyboard* (Middleton, WI: A–R Editions, 2006); Claude Cadoz, "Le Geste Canal De Communication Homme/Machine: La Communication Instrumentale," *Technique et Sciences Informatiques – Numéro Spécial: Interface Homme-Machine* 13, no. 1 (1994); Joseph Butch Rovan et al., "Instrumental Gesture Mapping Strategies as Expressivity Determinants in Computer Music Performance" (paper presented at the KANSEI: Technology of Emotion Conference, 1997).
4. Joseph Butch Rovan and Robert Wechsler, "Artistic Collaboration in an Interactive Dance and Music Performance Environment," in *Anomalie Digital Arts: Digital Performance*, ed. Emanuele Quinz (Paris: Les Editions HYX, 2002).
5. David Burroughs, "Instrumentalities," *The Journal of Musicology* 5, no. 1 (1987): 117.
6. Aden Evens, *Sound Ideas: Music, Machines, and Experience*, vol. 27, *Theory out of Bounds* (Minneapolis: University of Minnesota Press, 2005), 160–1.
7. Ibid., 167.
8. Ibid., 160.

9. Joel Ryan, "Some Remarks on Musical Instrument Design at Steim," *Contemporary Music Review* 6, no. 1 (1991): 7.
10. A notable exception is Ray Edgar's "Sweatstick," created at STEIM; *see* http://www.rayedgar.com/sweatstick.html
11. One conventional instrument that relies on a high degree of maintenance gestures is the pipe organ. The organist is constantly changing the configuration (registration) of the instrument through altering the pattern of stops. To the extent that most organists perform alone in a loft high above the audience, very few of these performance gestures are actually "expressive" in the sense used here.
12. In this case, exaggeration alters sound, since listeners tend to perceive causal relationships between what they see and what they hear. *See* Michel Chion, *Audio-Vision: Sound on Screen* (New York: Columbia University Press, 1994).
13. *See* www.soundidea.org for video documentation of *Continuities*.
14. *See* www.soundidea.org for video documentation of *COLLIDE*.
15. The software for *COLLIDE* comprises a set of real-time synthesis routines, a mutable sensor data-mapping scheme, and a separate software environment controlling interactive video. The "scenes" are defined by changes to the gesture mappings, synthesis algorithms, and algorithmic choices available to the video generator.
16. Other examples include Cléo Palacio-Quintin's extended flute, Matthew Burtner's extended saxophone, Curtis Bahn's extended electric double bass, or earlier instrument experiments like Gordon Mumma's extended French horn.
17. FSR pressure sensors are attached to the instrument, and capacitance touch sensors are rigged to several of the key assemblies. An accelerometer and gyroscopic sensor affixed to the bottom of the bell measure the horn's acceleration and rotation velocity. A separate control panel serves as the main interface for maintenance gestures; system states can be chosen, levels adjusted, algorithms started, and so on. The contact microphone and wireless audio transmitter send audio from performance gestures (key hits, taps on the horn) to the computer, which are used as additional sensor information or audio input to sound processing routines. The contact microphone also allows me to bow the metal components of the horn.
18. For images and more information about MiMICS, *see* www.soundidea.org

Select Bibliography

Allegue, Ludivine, Simon Jones, Baz Kershaw, and Angela Piccini, eds. *Practice-as-Research: In Performance and Screen*. Basingstoke: Palgrave Macmillan, 2009.

Arlander, Annette. "Some Conversations . . . in Various Spaces." In *Knowledge Is a Matter of Doing: Proceedings of the Symposium Theatre and Dance Artists Doing Research in Practice*, ed. Pentti Paavolainen and Anu Ala-Korpela, 118–23. Helsinki: Theatre Academy, 1995.

——. "Tila Ja Aika Eli Miten Megalomaanisesta Idealistista Tuli Suhteellisuudentajuinen Pragmaatikko (Space and Time – or How a Megalomaniac Idealist Turned into a Pragmatist with a Sense of Proportion)." In *Taide, Kertomus Ja Identiteetti (Art, Narrative and Identity)*, ed. Pia Houni and Pentti Paavolainen, 48–65. Helsinki: Theatre Academy, 1999.

Atkinson, Paul, et al, eds. *Handbook of Ethnography*. London, Thousand Oaks, CA, and New Delhi: Sage Publications, 2001.

Baily, John. "Ethnomusicology, Intermusability, and Performance Practice." In *The New (Ethno)Musicologies*, ed. Henry Stobart, 117–34. Lanhan, MD: Scarecrow Press, 2008.

Barrett, E., and B. Bolt, eds. *Practice as Research: Context, Method, Knowledge*: I. B. Tauris, 2007.

Béhague, Gerard, ed. *Performance Practice: Ethnomusicological Perspectives*. Westport, CT: Greenwood Press, 1984.

Borgdorff, Henk. *The Debate on Research in the Arts*. Bergen, Norway: Kunsthögskolen i Bergen, 2006.

Brydon-Miller, Mary, Patricia Maguire, and Alice McIntyre, eds. *Traveling Companions: Feminism, Teaching, and Action Research*. Westport, CT: Praeger Publishers, 2004.

Conquergood, Dwight. "Performance Studies: Interventions and Radical Research." *TDR: The Drama Review* 46, no. 2 (2002): 145–56.

Crease, Robert. *The Play of Nature: Experimentation and Performance*. Bloomington: Indiana University Press, 1993.

Dean, Roger, and Hazel Smith, eds. *Practice-Led Research/Research-Led Practice in the Creative Arts*. Edinburgh: Edinburgh University Press, forthcoming 2009.

Dixon, Steve. *Digital Performance: A History of New Media in Theater, Dance, Performance Art, and Installation*. Cambridge, MA: MIT Press, 2007.

Duxbury, L., E. M. Grierson, and D. Waite, eds. *Thinking through Practice: Art as Research in the Academy*. 2nd edn. Melbourne: The School of Art, RMIT University, 2008.

Edgar, I. *A Guide to Imagination-Based Research Methods*. New York and London: Routledge, 2004.

Elkins, James, ed. *Artists with PhDs: On the New Doctoral Degree in Studio Art*. New York: New Academia Publishing, 2009.

Greenwood, Davydd, and Morten Levin. *Introduction to Action Research: Social Research for Social Change*. London and Thousand Oaks, CA: Sage Publications, 2007.

Hannula, Mika, Juha Suoranta, and Tere Vadén. *Artistic Research – Theories, Methods and Practices*. Helsinki: Academy of Fine Arts; Gothenburg, Sweden University of Gothenburg/ArtMonitor, Gothenburg 2005.

Haseman, Brad. "A Manifesto for Performative Research." *Media International Australia Incorporating Culture and Policy* no. 118 (2006): 98–106.

——. "Rupture and Recognition: Identifying the Performative Research Paradigm." In *Practice as Research*, ed. E. Barrett and B. Bolt. London: I. B. Tauris, 2007.

——. "Tightrope Writing: Creative Writing Programs in the RQF Environment." *TEXT* 11, no. 1 (2007).

Hauptfleisch, Temple. "Artistic Outputs, Arts Research and the Rating of the Theatre Practitioner as Researcher – Some Responses to the NRF Rating System after the First Three Years." *South African Theatre Journal (SATJ)* 19, (2005): 9–34.

Hesse-Biber, Nagy, and Patricia Leavy, eds. *Approaches to Qualitative Research: A Reader on Theory and Practice*. New York and Oxford: Oxford University Press, 2004.

Hunter, Lynette. *Critiques of Knowing: Situated Textualities in Science, Computing and the Arts*. London: Routledge, 1999.

Jackson, Shannon. *Professing Performance: Theatre in the Academy from Philology to Performativity*. Cambridge: Cambridge University Press, 2004.

Janik, A. "Tacit Knowledge, Working Life and Scientific 'Method'." In *Knowledge, Skill, and Artificial Intelligence: Tacit Knowledge and New Technology*, ed. Bo Goranzon and Ingela Josefson, 53–63. London: Springer Verlag, 1987.

Kaye, Nick. *Site-Specific Art: Performance, Place and Documentation*. London and New York: Routledge, 2000.

Kershaw, Baz. *Theatre Ecology: Environments and Performance Events*. Cambridge: Cambridge University Press, 2007.

Kilijunen, Satu, and Mika Hannula, eds. *Artistic Research*. Helsinki: Fine Art Academy, 2002.

Knowles, J. Gary, and Ardra L. Cole, eds. *Handbook of the Arts in Qualitative Research*. Los Angeles: Sage Publications, 2008.

Lave, Jean, and Etienne Wenger. *Situated Learning: Legitimate Peripheral Participation*. Cambridge: Cambridge University Press, 1991.

Lewin, Kurt. "Action Research and Minority Problems." *Journal of Social Issues* 2, no. 4 (1946): 36–46.

MacLeod, Katy, and Lin Holdridge, eds. *Thinking through Art: Reflections on Art as Research*. London: Routledge, 2006.

Madison, D. Soyini. *Critical Ethnography: Method, Ethics, and Performance*. Thousand Oaks, CA: Sage Publications, 2005.

Mienczakowski, Jim. "Ethnodrama: Performed Research–Limitations and Potential." In *Handbook of Ethnography*, ed. Paul Atkinson et al., 468–76. London, Thousand Oaks, CA, and New Delhi: Sage Publications, 2001.

Paavolainen, Pentti, and Anu Ala-Korpela, eds. *Knowledge Is a Matter of Doing: Proceedings of the Symposium Theatre and Dance Artists Doing Research in Practice*, Helsinki: Acta Scenica 1, 1995.

Perks, Robert, and Alistair Thomson, eds. *The Oral History Reader*. 2nd edn. New York: Routledge, 2006.

Piccini, Angela. "An Historiographic Perspective on Practice as Research." *Studies in Theatre and Performance* 23, no. 3 (2003): 191–207.

Piccini, Angela, and Baz Kershaw. "Practice as Research in Performance: From Epistemology to Evaluation." *The Journal of Media Practice* 4, no. 2 (2003): 113–23.

Pickering, Andrew. *Science as Practice and Culture*. Chicago: University of Chicago Press, 1992.

Pollock, Della, ed. *Remembering: Oral History Performance*. New York: Palgrave, 2005.

Reason, Peter, and Hilary Bradbury, eds. *Handbook of Action Research: Participative Inquiry and Practice*. London and Thousand Oaks, CA: Sage Publications, 2008.
Rouhiainen, Leena, ed. *Ways of Knowing in Dance and Art*, Acta Scenica 19. Helsinki: Theatre Academy, 2007.
Stewart, R. "Practice Vs Praxis: Constructing Models for Practitioner-Based Research." *TEXT* 5, no. 2 (2001): 4.
Stringer, Ernest T. *Action Research*. 3rd edn. London and Thousand Oaks, CA: Sage Publications, 2007.
Sullivan, Graeme. *Art Practice as Research: Inquiry in the Visual Arts*. Thousand Oaks, CA: Sage Publications, 2005.
Taylor, Millie. "Schrödinger's Research: When Does Practice as Research Occur?" *South African Theatre Journal (SATJ)* 19, (2005): 183–90.

Organization websites

Arts and Humanities Research Council (AHRC), UK: www.ahrc.ac.uk
Imagining America: Artists and Scholars in Public Life, US: www.imaginingamerica.org
National Research Foundation (NRF), South Africa: www.nrf.ac.za
Organisation for Economic Co-Operation and Development (OECD): www.oecd.org
Research Assessment Exercise (RAE), UK: www.rae.ac.uk
Social Sciences and Humanities Research Council (SSHRC), Canada, www.sshrc.ca
The Centre for Performance Research (CPR), Wales: www.thecpr.org.uk

Webpages

AHRC. "Research Grants." http://www.ahrc.ac.uk/apply/research/research_grants.asp. Retrieved 30 August 2008..
Gray, C. "Inquiry through Practice: Developing Appropriate Research Strategies." http://www2.rgu.ac.uk/criad/cgpapers/ngnm/ngnm.html. Retrieved 12 January 2005.
Kershaw, Baz. "What Is Practice as Research?" Practice as Research in Performance, University of Bristol: http://www.bristol.ac.uk/parip/faq.htm#menu. Retrieved 5 May 2008.
Nelson, Robin, and Stuart Andrews. "Practise as Research: Regulations, Protocols and Guidelines: A Short Report..." Palatine: http://www.palatine.ac.uk/files/903.pdf. Retrieved 20 August 2008.
NRF. "Report on a Workshop on Research in Drama, Theatre and Performance in South African HE Institutions." UCT Drama Website: www.drama.uct.ac.za. Retrieved 5 May 2008.
——. "Definition of Rating Categories." http://evaluation.nrf.ac.za/Content/Documents/ Rating/ratingcategories.doc. Retrieved 8 May 2008.
——. "Definition of Research." http://evaluation.nrf.ac.za/Content/Documents/Rating/definition_research_2005.PDF. Retrieved 8 May 2008.
OECD. *Frascati Manual*. Fifth ed. 2002: http://europa.eu.int/estatref/info/sdds/en/rd/rd_frascati_manual_2002.pdf. Retrieved 3 January 2009.
RAE. "Criteria for Assessment 66 Drama, Dance and Performing Arts." HERO: http://www.hero.ac.uk/rae/rae96/66.html. Retrieved 30 August 2008.
——. "Section III: Panels' Criteria and Working Methods." HERO. http://www.hero.ac.uk/rae/Pubs/5_99/ByUoa/crit66.htm. Retrieved 30 August 2008.

——. "Definition of Research and Eligible Outputs 2008." http://www.rca.ac.uk/pages/research/definitions_of_research_3091.html. Retrieved 30 August 2008.

——. "UOA 65, Drama, Dance and Performing Arts." http://www.rae.ac.uk/pubs/2006/01/docs/o65.pdf. Retrieved 30 August 2008.

Richards, Alison. "Appendix A: Performance as Research/ Performance as Publication." http://www.adsa.edu.au/research/performance-as-research/appendix-a. Retrieved 1 November 2008.

——. "Discussion Paper: Performance as Research/ Research by Means of Performance." http://www.adsa.edu.au/research/performance-as-research/performance-as-research. Retrieved 21 September 2008.

Richards, Alison, and B. Dunstone. "Appendix B: Some Issues in Performance as Research." http://www.adsa.edu.au/research/performance-as-research/appendix-b. Retrieved 1 November 2008.

Rockwell, Geoffrey. "Research/Creation, Again." geoffreyrockwell.com. http://www.philosophi.ca/theoreti/wp-content/uploads/notes/001092.html. Retrieved 5 May 2008.

SSHRC. "Definitions." http://www.sshrc.ca/web/apply/background/definitions_e.asp. Retrieved 5 May 2008.

——. "Congress." http://www.fedcan.ca/congress2005/programs/6_1.htm. Retrieved 5 May 2008.

——. "Framing Our Direction." http://www.sshrc.ca/web/about/publications/framing_our_direction_e.pdf. Retrieved 5 May 2008.

Strand, Dennis. "Research in the Creative Arts: Executive Summary and Recommendations." Canberra School of Art, The Australian National University: http://www.dest.gov.au/archive/highered/eippubs/eip98-6/execsum.htm. Retrieved 21 February 2008.

Index